# VILLAGE WISDOM / *FUTURE CITIES*

# VILLAGE WISDOM

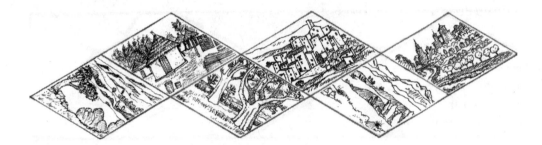

# *FUTURE CITIES*

## The Third International
## Ecocity and Ecovillage Conference
## held in Yoff, Senegal, January 8-12, 1996

**Conveners**
Serigne Mbaye Diene and Joan Bokaer

**Hosts in Senegal**
The Association for the Economic, Cultural
and Social Promotion of Yoff (APECSY),
the Mayor of Dakar, Mamadou Diop
and the traditional villages of Yoff, Ouakam, and Ngor

**Editors**
Richard Register and Brady Peeks

**Introductions**
Richard Register

Ecocity Builders
Oakland, California, USA

# Village Wisdom / Future Cities

Edited by Richard Register and Brady Peeks

Cover and book design by Bill Mastin
Front cover photo by Susan Felter
Back cover photos, top by Ray Bruman,
lower by Jeff Kenworthy
Illustrations credited beside all photos or art

Printed in the USA on acid-free recycled paper by Alonzo Press,
Hayward, California

Ecocity Builders is an Oakland, California-based educational
non-profit corporation exploring and promoting ecologically healthy
communities through design, planning, research, community
projects and production of educational events and materials. The
organization is the publisher of VILLAGE WISDOM/FUTURE CITIES –
THE THIRD INTERNATIONAL ECOCITY CONFERENCE and advised EcoVillage
at Ithaca and The Association for the Economic, Cultural and Social
Promotion of Yoff (APECSY), in Yoff, Senegal in the production of
that conference. For information on participating in the organization's
local and international projects and for information about future
International Ecocity Conferences, direct inquiries to Ecocity Builders,
5427 Telegraph Ave., W2, Oakland, CA 94609, USA.

Library of Congress Catalog Card Number: 97-60014

VILLAGE WISDOM / FUTURE CITIES – THE THIRD INTERNATIONAL
ECOCITY AND ECOVILLAGE CONFERENCE
Edited by Richard Register and Brady Peeks
ISBN 0-9655732-0-6
1. – Urban ecology.  2. – Environmental policy.  3. – Architecture.
4. – Design.  5. – Agriculture, urban.  6. – City planning and policy.
7. – Transportation, urban.  8. – Village culture and traditions.
9. – Africa

Conference logo
by Lisa Cowden

Yoff, Senegal.

*To those so deep in the past,*
*their bones are city ruins*
*and their ghosts are the genes within us,*

*To those so deep in the future,*
*all children become our children*
*and Earth's village a home for all species,*

*To all those, that we may remember*
*and never cease dreaming and building,*
*in timeless respect of our giving.*

# Table of Contents

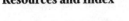

# Acknowledgements

This book about a long journey and many days of meetings was itself a long journey and many days, alas, many *months* of meetings.  But we weathered it all and now we have the book you see in your hands representing the Third International Ecocity Conference, which for most of you happened in a place far, far away.  VILLAGE WISDOM / FUTURE CITIES would not have seen the light of day if not for the assistance of a large number of people contributing, from a few hours to hundreds of hours of their time, labor, skills and imagination.

Thanks everyone, and especially to Brady Peeks who was there through several intensive work sessions of two to three weeks each, editing, keeping the computers filling and formatting.  He provided assistance of a wide range throughout the process of editing and publishing this book.

Special thanks to Bill Mastin, on board as book designer, and photographer.  Bill's wife, Susan Felter, graced these pages with her photography and computer skills. Thanks to Santa Clara University's Art Department and Department chair Kelly Detweiler for computer equipment that helped produce the book.  Sumner Stone donated his elegant Stone typefaces, used throughout.

Supporting us were members of our editorial committee: Joan Bokaer, the conference co-convener, Peter Harnik, one of our presenters at the First International Ecocity Conference in 1990, Chris Canfield, editor of the report on the First International Ecocity Conference,  and Mignon O'Young, Nathan Lozier, Patricia Zuker, Stacy Giotto, Charlotte Rieger, Dione Early and Ondine Wilhelm.  Thanks for translation, French and Russian to English, to Paule and Pierre Aubery, Dixie LaGrande, Alanso Mancos, Christine Menagér, Mike and Sylvie Rubenstein and Elena Bridgman.

Special thanks to Karil Daniels, without whose work much of this book's contents simply would not have been available – almost half of the words in these pages are from transcriptions from her seventeen hours of video tape. Thanks to Kate Dorman for editing a short video on the conference which helped the editorial team visualize the book, and to Chris Canote, for video capture of Karil's imagery.

Tim Hansen, Ecocity Builders' Treasurer, and Linda Levitsky helped greatly with planning, and Betsy DuVal, Ben Murphey,Elizabeth Hennin, Brian Block, Ray Bruman, Marco Marquez, Gale Mitchell and Nancy Lieblich  helped with a variety of jobs during editing and production.

We were able to publish VILLAGE WISDOM / FUTURE CITIES due to a loan from Gaia Trust of Denmark, and gifts from Charlotte Rieger, Pat and Graeme Welch, Nancy Lieblich, Michael Vandeman and other Ecocity Builders members.

Cerro Gordo, Oregon – notably Brady Peeks, Chris Canfield and Christine Menagér – helped by providing office space and other support in their community during the crucial period of launching work on the publication.

Of course we owe most of all to the conveners of The Third International Ecocity Conference itself.  Those were the courageous two people who took the risks and put forward the enormous effort to gather hundreds of people from all the inhabited continents of Earth to a place that seems remote to most, yet right in the center of the crucial issues humanity faces today.  They were Serigne Mbaye Diene and Joan Bokaer.  Behind Serigne, the entire village of Yoff, with more support from the villages of Oukam and Ngor, lent enormous energy and kind generosity to the enterprise.  The Mayor of Dakar, Momadou Diop, rented the conference center for us and provided busses and essential secretarial services.  The Minister of the Environment and Protector of Nature, Abdoulaye Bathily, and President of Senegal, Abdou Dioff, lent their assistance and crucial endorsement.

The conference planning committee included Liz Walker, Tukumbi Lumumba-Kasango, Marci Riseman, Remileku Rakiatu Cole, Abdou Azziz, Stevie Chinitz and Mary Webber, financially assisted by Marci Riseman, Alan Day, Katie Branch and Ecocity Builders.  The International Ecocity Conference Relay Committee was made up of myself, Paul Downton, Cherié Hoyle, Karl Linn, Bill Mastin, Charlotte Rieger, Arthur Monroe, Nancy Lieblich, Susan Felter, Jon Li, Michael Warburton, Eva Queye and Sylvia McLaughlin.

My heartfelt thanks to all,        Richard Register

# Preface

## RICHARD REGISTER

Imagine the immense Earth turning beneath our view, huge continents appearing on one horizon, rolling to the other. Below, endless blue seas, glorious mountains, forests and grasslands march into and retreat from our view. Shadows stretch, and with a flash of sunset, twilight turns to dizzying countless stars above and equally as many life forms below, each one, like us people, seeking its place, trying to make a living.

This world is seething with life, sparkling with cities and towns lit up at night, fuming and clattering by day, growing in number, spreading wider, driving out our natural companions on this Earth, all the way to extinction, changing the climate, exhausting resources both mineral and chemical, biological and spiritual – and poisoning most of the rest. It need not be so.

Here on the edge of millennia, our biosphere is deep in crisis – a crisis that was confronted head-on with enormous creativity and vision by the Third International Ecocity Conference.

On January 8, 1996, up went the black starry curtain and under poured in the sunrise. Africa rose up over the Atlantic horizon and the conference commenced. For five days our little jet-powered aluminum tubes had been landing here, like tiny insects, on the edge of the great desert and endless sea. About 120 of us, some from every inhabited continent, mostly people from wealthy countries, had been debarking from the planes, joining another 180 of us, mostly from Senegal and other African countries. We came to honor one another's efforts to honor the Earth, its people and all attempts to build our communities in balance with nature. The wisdom of the villager, almost always overlooked in the rush to material development and "success," came face to face, person to person, with the courage of the urban ecocity innovator,

*Richard Register is chief editor of this book, author of the upcoming ECOCITIES, President of Ecocity Builders, Inc. of Oakland, California, and convener of the First International Ecocity Conference.*

Richard Register

The beach at Yoff.

often struggling, with pennies and deep convictions, to build a healthy future.

We lived with the people, slept in their homes, ate the produce of their fields and the catch from the huge waves to the north, west and south of the Cap Vert peninsula. We breathed in fish-scented salt water air and tasted the dusty desert breezes. We compared cities and towns of all sizes on every continent, shared our efforts at reshaping, recycling, composting and enhancing biodiversity. We launched into rethinking agriculture, fisheries, architecture, technologies, planning and design. We scrutinized the past and future, asking what it was people had done and should do next to create a vital culture and preserve nature in all its enormity and subtlety. We left a little wiser for the effort to understand, and certainly, enriched by the culture exchanged, the friendships made, the dreaming and scheming accomplished together. Back home we began our work again, in our towns, our organizations, among our friends, families and colleagues with many new stories and a few new tools.

Now, in this ambitious small book, we can share the event with friends around the world. Will these pages and words change your life? Will they make many more of us believe we can change the way we build and live? Will they give us new ways to stop the destruction of the natural world and get on with building a truly creative, compassionate civilization right at the time its life support systems are beginning to crumble in our hands?

The organizers of this conference set out to explore the wisdom of the traditional village. They wanted to share the best efforts from around our home planet in a place where the ideas could be reinforced by powerful personal experiences. Without any question, in this they attained their goal. They wanted to make unexpected discoveries in the mix of races, histories, religions, languages, and conditions of material wealth and social security from both urban and rural environments, and as I think you will see, largely succeeded here too.

Now, with the conference reconvened in this book – take two, this time with you, dear reader, in the cast – we participate again in the growth of a movement for ecologically healthy communities. The Third International Ecocity and Ecovillage Conference was one of those rare events capable of echoing deep into the future. May it reach into your lives as it has reached into the lives of all of us who went there and lived there through those fine African days and nights.

Turning an international conference into a book is a true editorial challenge, even for an event like Ecocity 3 with its many charismatic characters, dramatic confrontations with profound problems and its spectacular location. Conference reports aren't supposed to be exciting. Well, this one might be, because that's the way the event was. If we can draw you into the experience of being there, you will find a real book here and not the usual proceedings of drowsy repute.

Convening a conference is challenging enough. It's art more than science, poetry more than permutations, village and tribal more than city and family. It's something like writing a play and giving the actors not a script but a schedule of appearances. You expect them to say exactly what they think is most important about their work in relation to the conference themes. End expectations here: they get to write the script.

In addition, the cast gets to live the story – literally village-style , at our conference – for a heightened small slice of their lives, in the freedom to take the event wherever they and the vicissitudes of fate collectively decide. Neither a highly predictable artistic medium, this, nor a tightly controlled laboratory experiment. In the case of the ecocity conferences, the first held in Berkeley, California in 1990 and the second held in Adelaide, Australia, 1992, the tradition was already well established that the event was to be a real community conference, featuring as many citizen activists, as professionals, as academics, as government people. This too contributed to the freshness, and the decidedly un-stuffiness of the tradition. Among conferences, these have been even less scripted than usual.

Problems with translation, cross-cultural confusion, having to slow down speech in order to be understood and needing to concentrate extra hard to understand accents – all these are problems that plague international conferences, but add enormously at the same time. The lesson is that the effort is worth it. What you might miss in terms of precision you are likely to gain in exercise of your own imagination and development of intuitive grasp.

In this book we try to give you the feeling of being there. If words like "centre," "organise" and "whilst" throw the American English speaker, understand that the Australians, English and English-speaking Europeans find our (this is being published in the United States) equivalents similarly peculiar. Words in Wolof, the native language of most of Senegal, may be hard to pronounce, but try those speakers' and places' names anyway, like the participants in the conference did; pronounce them slowly and soon you begin to feel the place and the human exchange in a way you never will without the effort. Being a little on edge and a little adrift, experiencing an uncomfortable degree of ambiguity – then sheer acceptance – *is* the experience of such conferences, especially this one with its participants' extremely broad-ranging kinds of work and conditions of life. Be patient – these are real people, self-taught and highly educated, from down on the farm and in the streets to highly placed

Richard Register

Small ecological city surounded by natural and agricultural land in a typical northern California environment.

in government office and academic post. Mostly their words are lightly edited to preserve the way imperfect grand efforts are. Reciprocate, even if your eyes passing over these pages are distant from the village of Yoff, the *teranga*, or traditional warm hospitality of Senegal.

Know too that we the editors and the translators were struggling for meanings and trying to faithfully render ideas and nuances – just as were the conferees. Finally, we will do our best to share the midnight drum beat reverberating down the narrow streets on carrier waves of distant ocean surf. But for some things, you just had to be there. There is always Ecocity 4!

A word about the conference name: the first two ecocity conferences did not mention "ecovillage" in the conference title, and were informally known simply as Ecocity 1 and Ecocity 2. The villagers of Yoff proudly renamed #3 on their own. When the visitors arrived from around the world, there were signs printed on cloth banners over the narrow streets, painted on the walls and even lovingly lettered in bright colors on the sides of boats in the village fishing fleet, and they all welcomed us to La Troisieme Conference Internationale des EcoVilles et EcoVillages. Organizers of the first two conferences had assumed that the term "ecocity" was inclusive of all scales of organization – city, town, village, quarter, neighborhood – of what perhaps would more properly be called the "ecommunity." In this book, we defer to our hosts in the title, and let all our speakers and readers decide the best names, spellings, use of capitals and so on on their own.

A word about organization of this book. We are exploring whole systems thinking here, piecing together the complex body and spirit of the city, the town, the village, in both its physical and its human dimensions, its form and its function, its anatomy and its metabolism. This meant that many speakers were talking about two or more things at once. Try to organize *that* into simple sequential categories! Well, we did the best we could, trying to draw out certain themes we felt were important and unifying. Only time and you the reader can tell if we made the best choices in this regard.

Warning: these are mostly synopses of talks, not full papers. Some pieces here were translated from French to English from video tapes, then synopsized, or in some cases, if shorter than deserved, supplemented with more material from or about the particular presenter. We have been as faithful to the content and intent as we could, reporting on a conference in three languages with small financial support.

A comment on that too, since it was so much a part of the experience. The host village of Yoff is called "poor" by many. And their material resources, if not their cultural resources, were genuinely stretched to the limit. So too for many of the conferees. Hosting the conference and attending it, for almost all of us, was a labor of love – and many of us will be paying back loans for months because we attended. There was some substantial support from the office of the Mayor of Dakar, Mamadou Diof, who rented the conference center for us, but almost none from the institutions that often sponsor conferences. And so, the spirit of the event was this: our civilization, and the whole Earth because of it, is in crisis. But we looked into one another's eyes across a wide canyon of cultural, religious, racial and economic differences – face to face just inches apart – and, assessing our modest powers and immense problems, confronted both crisis and wonderful new possibilities together. Together we will build a better future.

# Greetings from the Second International Ecocity Conference Convener

## PAUL DOWNTON

Paul Downton
in the streets of
Palermo, Italy.

Greetings fellow frogs!

That "g'day" to all us endangered blokes was the call from Down Under to the First International Ecological City Conference. It was inspired by the urban myth made famous by David Suzuki. I gave one of the plenary talks at EcoCity 1 and called on the conference participants to *escape from the city of boiling frogs!* The myth is that a frog in a pot of water being slowly heated, does not notice the gradual change in temperature, and so, is boiled alive. I proposed that we compare the conditions in our cities with the mythical frogs in their pots, and resolve that there had to be a way out – or we were headed for a similar fate.

The imagination, enthusiasm, commitment and sheer joy of that first conference, convened by eco-city pioneer Richard Register and his team of urban ecologists, carried half-way around the globe and two years later, Adelaide's own urban ecologists convened the Second International EcoCity Conference. In the tradition of EcoCity 1, the members of the community non-profit Urban Ecology Australia committed that conference to practical outcomes for a positive and empowering impact on the community, carrying the seed of ecocity activism into the mainstream of urban life.

EcoCity 1 saw the bringing together and coming of age of a whole new global movement. EcoCity 2 saw the birth of the first full-scale live project to bring together the principles and vision which inspire that movement. Designed and developed by the Ecopolis Design Team, in which I participated throughout, that project is Wirranendi – the Halifax EcoCity Project, and it was selected as part of the Best Practices Exhibition at the Second United Nations Conference on Human Settlements, Habitat II, held in Istanbul, Turkey in June, 1996.

*Paul Downton is Professor of Architecture at the University of South Australia in Adelaide, Australia and co-convener, with Chérie Hoyle, of the Second International Ecocity Conference, held in Adelaide in 1992. He is also co-founder with Ms. Hoyle of Urban Ecology Australia, Inc.*

Working from the Centre for Urban Ecology, co-ordinated by Chérie Hoyle, hundreds of volunteers, mostly young people, have, over more than four years, created a practical, working model for community-driven ecological urban redevelopment based on principles of social justice and ecological restoration. That model, born in the crucible of community action, raised in the glare of media attention and the criticism of entrenched interest groups, has affected government at the national, regional and local level. It has influenced the thinking of industry. It has become an icon for popular, progressive change towards an ecologically and socially sustaining future.

Wirranendi, the EcoCity Project on Halifax Street, has consumed the energy and resources of Urban Ecology Australia as we pursued the goal of making our vision come alive. But as conveners of EcoCity 2, we joined with Richard Register, convener of EcoCity 1 and most of his original Conference Planning Committee, to review the proposals and locales for EcoCity 3. We wanted to see the movement given voice in a developing country. We had no hesitation in supporting the bid to host EcoCity 3 in Senegal.

As we try and greet the uncertain future of a world where a billion people lack electricity, two billion people lack decent shelter and god knows how many are starving or diseased, the greatest task ahead of us is to rebuild the city, that greatest of human endeavours. The city is the centre of civilisation, hub of commerce and culture, and the most powerful tool we have ever found for resource management and human advancement.

The eco-city is the next step in the evolution of our urban environments: built to fit its place, in co-operation with nature rather than in conflict; designed for people to live whilst keeping the cycles of atmosphere, water, nutrients and biology in healthy balance; empowering the powerless, getting food to the hungry and shelter to the homeless; creating a place for everyone, in every land, for all time.

In Australia we have struggled to bring this vision into living existence through the power of a united community. We know that ordinary people can make a difference and that ideas can be turned into action. So for all you frogs seeking escape from the turmoil of troubled waters, there is a way out. The practical, realisable, empowering vision of ecological cities is the means by which people around the world can create their own future habitat, making new cities, remaking old cities, and healing the planet.

The message of greeting from all of us in the Centre for Urban Ecology, home of Wirranendi – the Halifax EcoCity Project, is simple: escape from the cities of boiling frogs!

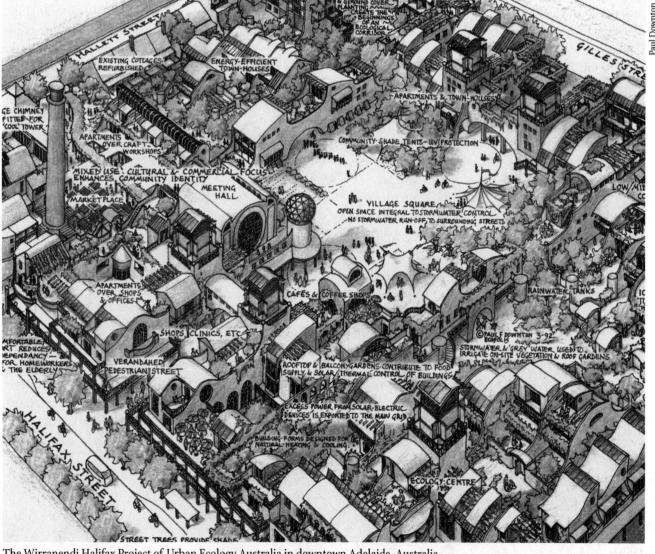

The Wirranendi Halifax Project of Urban Ecology Australia in downtown Adelaide, Australia.

Karil Daniels

*Throughout the conference, participants were treated to the traditional hospitality, known locally as* terenga, *with home lodging, meals, dance performances, open participation in traditional ceremonies, meetings with elders and tours to innumerable sites, both historic and grand, and typical and humble.*

Above, drummers and dancers at the opening ceremonies of the conference.  Music and dance bring together young and old and connect ancient traditions with current village life in Yoff, Ngor and Oukam.  Below and left of cener,  Liz Walker, conference speaker and co-director of EcoVillage at Ithaca, dances at our welcome reception hosted by Yoff villagers.

Karil Daniels

# Save the Village, Restore the Earth... Start the Conference

## JOAN BOKAER & SERIGNE MBAYE DIENE

**Joan's Greeting:**

First of all I would like to thank the President of Senegal, Abdou Dioff who is also the honorary president of our conference. (*Mr. Dioff was out of the country attending the funeral of François Mitterrand and could not be present.*) I would like to thank Prof. Abdoulaye Bathily, Minister of the Environment who is known in places as far away as Cornell University, where I work, as a very great thinker. And a very special thank you with a great deal of warmth and affection to the Mayor of Dakar, who has done so much to make sure this conference will be a success, who came to visit us in Ithaca more than a year ago to talk with us about the conference.

(*Joan continues thanking everyone who has come from afar, mayors and members of city governments, people in attendance from the traditional villages near Dakar, and...*) the APECSY members who have worked so hard to make all of us feel so welcome, and to all the people of Yoff.

There is one thing we all know for sure that we have in common now, and that is after just a few days in your traditional village we all feel like family. (*loud clapping*)

Do you realize the significance of this? (*louder clapping*) That people can come from thirty different countries starting out as total strangers, and then we discover we are not strangers at all; we all feel at home.

My greatest hope is that we will be so successful in our work here, in what we are striving to do, that one day all of you will be able to come to our countries and instantly feel at home. (*Loudest clapping so far.*)

I am going to talk briefly about the people who have come here to Yoff specifically to honor the traditional villages. Some people may feel awkward in this large, gracious hall, but I will have people stand up anyway, so

*Joan Bokaer is founder and co-director of EcoVillage at Ithaca, Ithaca, New York and co-convener of the Third International Conference on Ecocities and Ecovillages. Here she is speaking before approximately 1,000 people on the opening day of the conference, with three-quarters of the seats occupied by villagers from the region surrounding Dakar.*

APECSY headquarters in Yoff, Senegal.

that we can all see those from far, far away. (*She continues by introducing all the presenters and many of the participants, whom you will meet later in these pages, continent by continent, person by person.*)

This gathering represents a variety of backgrounds. We have architects, planners, geographers, anthropologists, political scientists, teachers, students, citizen activists, farmers. We have the builders, the growers and the makers. Because the ecological city concept is an integrative one. And we are now seeking to synthesize the old with the new. We have come here because we feel we have a lot to learn from you. Together during the next four days we will be meeting and talking and listening to one another and helping to clarify what it is we want, what kind of world we want, because in this room we have the people it takes to change the course of humanity.

I am going to finish by introducing a very special person. Serigne Mbaye Diene, please stand up. He comes from a strong background of traditional thinking. (*Loud clapping.*) I want to tell you what happened. What happened was that somebody rooted in the traditional village came to a very high-powered major international University. This is a man who understands his roots, and yet his dissertation, which was on nutrition and delivery of health services, was considered brilliant and is changing the way people think in his field in that University, and perhaps soon, in academic circles around the world.

**Serigne's Greeting:**

Mr. Minister of the Environment and the Protection of Nature, Mr. Mayor of Dakar, President of the Urban Committee of the Commune of Dakar, ladies and gentlemen, members of the Diplomatic Corps, Dignitaries of the Lebou Collective, Djarafs, Ndey Dji Reews, Saltiques, Jambirs and Freys (*administrators and decision makers of the village councils whom you will read about presently*) Mr. Grand Serigne of Dakar, honorable participants, dear guests:

First of all, I wish to welcome you to Dakar, all you participants in the Third International Conference of Ecovillages and Ecocities that opens today.

As you know, the central theme of the conference is "to integrate traditional African village wisdom into an ecological reconstruction for the future, a reconstruction of our cities and villages in balance with nature."

This is to say that the villages and traditional quarters of Dakar that were chosen to host and shelter the conference have the difficult privilege and delicate mission to represent the continent of Africa in this noble enterprise of ecological rebuilding.

This choice was not happenstance, because the traditional villages of Dakar, disposed all along the coast of this most western part of Africa, subsists daily with the perverse effects of the multiple aggressions of growth and sprawl of Dakar City, eroding their lifestyle, economy and culture. But at the foundation of their wisdom, the solidity of their centuries-old traditional social and political institutions have been able to maintain and preserve essential principles and fundamental social and religious traditions. I only have to offer as proof the annual tribute of Yoff Village to the spirit protector

---

*In 1995 Mr. Serigne Mbaye Diene, co-convenor of Ecocity 3, received his doctorate in nutrition from Cornell University in Ithaca, New York. He is the founder and current head of the Association for the Economic, Cultural and Social Promotion of Yoff and is currently working on health and nutrition delivery systems for eight African nations.*

Mame Ndaire, which many of you were able to participate in this morning.

Those of you who visited the village of Ngor yesterday were able to sit in the shade of a large tree 600 years old that is the jealous guardian of the village entryway, that the villagers have maintained and venerated with the same respect as the ancients did.

But there is a grave question that comes up today: For how much longer can the traditional people resist the intense and multiple cultural aggressions transported by means of mass communication controlled by the west? We do not have an answer to this question, but we have a deep and intuitive conviction that application of the program of ecological reconstruction that seeks to create an ecocity and ecovillage movement will be a good start in this direction. The uniqueness of this conference that meets today is in its conceptual approach and its method.

The organization of this conference was the result of a partnership between two organizations twinned together for three years on two different continents: EcoVillage at Ithaca in New York State in the United States, that is directed by Ms. Joan Bokaer and Ms. Elizabeth Walker, and the Association for the Economic, Cultural and Social Promotion of Yoff (APECSY) in Senegal, over which we have the honor to preside.

Joan, you had the idea to host an international conference on ecovillages and ecocities during a stay of a delegation from your organization visiting in Yoff in 1992. And we collaborated with you in your country. But if we have together been able to assemble in this room this magnificent group of eminent specialists and actors in the ecological domain representing thirty countries of all continents, it is thanks to the determination, perseverance and efficiency of your effort. We warmly thank you.

This is a part of other efforts on your behalf. Your dream to construct an ecological city model linked with rural agricultural villages, as a result of your historic nine month walk from the Pacific on the California west coast to the Atlantic on the New York east coast of your country has become a reality: EcoVillage at Ithaca exists; it is under construction at this moment. The reality of EcoVillage at Ithaca is like a honey jar (*Serigne holds up the gift of honey given to him by Jen and John Bokaer-Smith*) which came into existence through the heightened efforts of bees that two members of EcoVillage at Ithaca nourished by cultivating the flowers of Ithaca's gardens.

Our two organizations were joined in the preparation of this conference by the Association of Humanistic Culture of Yoff and Longjumeau in Longjumeau, France. Permit me to acknowledge the presence of Mme. Chantale Dauphin and of Claude Johon who directs this organization. We thank you for being here.

In addition to these organizations, another partner in organizing this summit was the Commune of Dakar. Mr. Mayor (*Mamadou Diop*), you have early on understood the intent and importance of Dakar City in welcoming such an avant-garde demonstration. You and your organization, in agreeing to be partners in organizing this conference, have at once demonstrated your open spirit and your willingness to trust and to associate the people and the managers of your city with the conference.

Mr. Minister of the Environment and Protection of Nature (*Abdoulaye Bathily*), the State of Senegal is not to be left out in the supporting preparations of this conference. Once he was informed, Mr. President of the Republic, Abdou Diouf, accepted the position of Honorary President of this conference. At that time he asked you to coordinate the work of a preparatory committee at the governmental level. We have appreciated the considerable efforts furnished by your department to accomplish this task and we express here our gratitude to the Chief of State together with his Government.

With respect to the method, three important roles are being played in the production of this conference. The traditional village population of Yoff invited most of the conference participants into their homes, providing a rare opportunity to share in the villagers' daily lives. Ngor and Oakum invited the conference participants to many events in their villages. The guests could then be in the best position to comprehend the villagers' reality and be able to exchange information and understanding effectively with them. Second, the conference organizers put together the concept of the program with its scientific and practical aspects and collected funds to finance most of the preparatory activities. And third, many of the organizers and villagers participated in workshops and as presenters in plenary sessions.

You will note organized cultural sessions throughout the conference, visiting mosques, schools, cooperatives, workshops, festivals and meals together and we would like to thank those providing these events. And our sincere thanks to UNESCO, le Grand Serigne and El Hadji Bassirou Diagne for financial support.

Ladies and gentlemen, we formulate a wish that a chain of solidarity for survival of the human species, of which we are a link, strengthen and maintain always.

Richard Register

Joan and Serigne.

Yoff, Sengal.

*An aerial photo of Yoff being studied at the workshop on design of the Yoff extension. The historic pattern of narrow pedestrian streets is clear, with the Atlantic coast to the left and a runway of Senegal's international airport, bottom right. In the last ten years, the open agricultural fields in the aerial view have been mostly filled in with housing. At the upper right, the Yoff extension, discussed in the Outcomes section of this book, is now being built.*

# 1. Where and When Are We, Anyway?

Where in the world are we and when in the march of time? We have arrived in a city on the westernmost tip of Africa. When the Lebou people settled here 500 years earlier, they began clearing the forest, then planting. Whatever was growing in the early colonial days, it was green enough to deserve the name Cap Vert – Green Cape. Today it looks nearly desert, sandy yellow brown and rusty red. Who was responsible? Or was it the one thousand years of encroachment of the Sahara, a "natural" phenomenon? The African flora and fauna now associated with East Africa, with elephants, lions, buffalo, wildebeest and giraffe, as depicted in ancient cave paintings in the desert, retreated from Senegal before the spreading sands.

In Yoff the population has more than doubled in the last twenty years and during that period most of the village's farm land became an airport, soccer stadium, cemetery, lots of private houses – and our conference center. Fish catches are way down and in places on the coast of Senegal, National Geographic reports, big European fishing ships pick up the small Senegalese vessels with cranes, lift the boats onto the "mother" ships' decks, travel as far away as 200 miles in search of fish, drop the small boats in the water to fish, pick them up again and take them back to point of origin – for fish at depressingly low prices.

Thus increasing numbers of people confront decreasing resources. Meantime, they have 40,000 people in Yoff and not a single police officer! How long can their peaceful ways last with outside media and commercial pressure compounding basic survival problems?

As far as history, current events and the city of Dakar are concerned, we are about to hear from the man who knows perhaps better than anyone else: the Mayor. Then a real expert on Africa will place ecocities in perspective with African history and economics. Finally, from my own experience in helping organize ecocity conferences and in attempting to further the movement for ecological building, I will give the best very short history I can muster on the deepest strains of the ecocity movement – starting way back even before beginnings of urban development about 9,000 years ago.

The poetic vision of Mayor Mamadou Diop will firmly locate us in this country. And the perspectives of Prof. Tukumbi Lumumba-Kasango will reveal our place on this continent. The ghosts of exploitation, slavery and injustice he introduces early on will be background for the whole conference, prodding us toward future solutions, shaping us as we try to reshape our communities. What I present says we, like the other creatures on Earth, have every right to change the world, so long as the change is healthy, so long as we understand we are participating in forces larger than ourselves and as ancient and future as all time.

Here in a place whose reality is a symbol for both injustice and hope, in a time when the last great land mammals of the world are on the run and the turning of the millennia promises *something* different, *very* different, we have some idea of what the best of it could be.

# Greetings from the Mayor of Dakar and Opening of the Conference

## MAMADOU DIOP

Mr. Minister of the Environment and Protector of Nature, honorable Council Members of Dakar, distinguished delegates and participants, honorable representatives of the Region of Dakar, Mr. President of the Association for the Economic, Cultural and Social Promotion of Yoff and the Cultural and Humanitarian Association of Yoff/Longjumeau of France, Mr. and Mrs. representatives of our local organizations, Citizens of Dakar, Villagers of Yoff, Ngor and Oukam, and dear guests:

It is a pleasure to welcome you to Dakar and to open the Third International Conference on Ecocities and Ecovillages.

The choice of Dakar and Yoff as site for this conference is a great honor for us because of the true importance of your ideals for humanity and our future. On behalf of the Municipal Council of Dakar I would like to extend our most sincere thanks and welcome.

Taking a panoramic view of the world, we have to face the reality that we are living in a time of great perils. But also I see auspicious signs of the times. There is so much disorder, so much destruction. The human species is becoming very vulnerable because of its assaults on the environment and the challenge to solve ecological problems has become one of the most important issues at the end of the millennium.

The fast growth of cities, especially in countries of the South, demands our immediate attention. The experts warn us that in the year 2000 half of humanity will be living in cities. But cities have grave problems, as the evolution of Dakar confirms. More than ever before, we face a decisive juncture. Dakar was born and developed on the peninsula of Cap Vert. It was on this thin strip of land that our city built its renown and it is here that it is paying the price of its success.

The city and its suburbs cover 525 square kilometers (*about 200 square miles*) at a density of 3,000 inhabitants per square kilometer (*about 7,100 per square mile*) for a total population estimated at approximately 1.6 million.

Internal growth and rural exodus are creating demographic pressures toward overpopulation. Thirty thousand new citizens arrive in Dakar every year. Our high rate of urbanization is aggravated by the economic crisis, especially the recent devaluation of the franc CFA (*the currency in several West African countries, including Senegal, reduced in value 50% in 1995*). The poor settle around the periphery of the capitol. In the move, these people are more de-ruralized than urbanized.

Because of the configuration of a site surrounded by ocean on three sides, the demographic push results in dwindling available space. As a result of this concentration of population and its material production, sanitation, health and habitat all suffer.

In addition, some behaviors of the people contribute to the problems. The city should cease to be the grounds for competition between different groups eager to develop their hegemony in the urban space which they want to organize and monopolize for their own particular interests.

The relations between the city and its suburbs is not harmonious either; the city aggressively sends out its tentacles upon its traditional villages and surrounding land to asphyxiate and consume them.

In this process we see in the fine details, the filigrane, the tragic shock and the brittle rupture between modernity and traditional values. We need to remember that progress, whatever it may be, must be in the service of humanity instead of taking humanity hostage. And so it is important for us to mobilize all our forces for life in search of solutions to these problems that confront us.

But now I think the time has come to debate the preservation of the environment in a new way, outside and beyond the usual circle of politics and administration. The time has come to call upon all those who live in our environment. You will understand that I appreciate the innovations that will be introduced by the Third International Ecocity Conference and I can be sure our visitors will appreciate their hosts, the population of Yoff, Ngor and Oukam, and their opportunity to discover the richness and diversity of our cultural heritage.

---

*Mamadou Diop is Mayor of Dakar and one of the hosts and patrons of Ecocity 3.*

Karil Daniels

Mamadou Diop

You will understand that the cultural and economic identity is for the village and the city what the genetic heritage is for humanity. Thus the necessity not to lose sight of the cultural dimension of the management of the city. Allow me to salute the relevance of the upcoming meetings which will facilitate a great moment of shared fertile exchange for the bringing closer of people, the betterment of international relationships and the reinforcing of peace in the world.

Ladies and gentlemen, in Senegal, the President, Abdou Diouf, understood very early on that the lack of mastering environmental problems constituted a very heavy handicap in the search and means to ensure the development of our country and the well-being of its people. It was for this reason that the Minister for the Environment and Protection of Nature was created and given to professor Abdoulaye Bathily, whose presence among us here today I am very pleased to announce, and whose remarkable direction and vision I do commend.

Ladies and gentlemen, members of the Third International Ecocity Conference, the fact that you are here confronting these central themes of your conference on environmental control and value proves that your effort is synchronized with its times. Of course your theme is very relevant and so too your inspiration which guided you, to Senegal and Dakar. Here you can always count on the support of the city of Dakar. It is not easy to assemble a meeting like this. Allow me to warmly congratulate the conference organizers.

Yesterday it was in Berkeley in California, in 1990. Then in Adelaide in Australia in 1992. Today the flame is rekindled in Dakar, city of commercial exchange, place of convergence of cultural Alizes (*the regional name for soft winds off the Atlantic and for cultural allies that border the Atlantic Ocean*), point of departure for humanity on its journey to conquest and destiny at the price of one thousand laments and one thousand eurekas.

From the tongue of Barbarie (*through which the ancestors of the native inhabitants of the Dakar region migrated from the opposite end of Africa 600 years ago*) to Saint-Louis (*the colonial capital of French West Africa*) up to the River of the South (*which flows from deep in the continent*), the Atlantic incubated and sheltered the living communities with the sea. The ocean is a fundamental element of the collective unconscious and is present in the mythology, science and symbols of Mame Ndiaré, Yumbur Yata, Gorgui Bassé (*the village spirits*) who guard and protect the blessed land of Yoff, Ngor and Oukam.

You said environment and culture. Let me tell you that I saw a dove, carrier of hopes, who left this country when one speaks of Indian summer, confided in me that she didn't promote hatred or revenge even if her peregrination retraced the tragic triangle of the slave trade and those trials during which the destiny of Young America, Old Europe and Mother Africa got knotted together forever. On the Beach of Yoff she looked at herself in fear of having changed. She saw that despite such a long journey across the ocean, the breeze of the Atlantic caressed her plumage, made it more bright, white like snow, white like the foam at her feet.

The patriarch passing by said that it was a good omen. So may the memory of the elders resuscitate and be the protector of your spirits and bless your presence.

Dal Leen ak Jaame. Thank you for your attention.

(*At this moment, in the audience of one thousand or more conference participants and others attending the opening ceremonies, a large block of villagers rose to their feet, and, to drum beat and swaying, began singing their praises for the Mayor and our gathering. Momentarily their griot (silent "t"), the keeper of their village oral history – six hundred years or more of their oral history – strode to the stage and sang several more praises, then sat down, followed by enthusiastic clapping. Then silent attention waiting for the next talk.*)

# Reflections on EcoCity Systems: Historic and Theoretic Perspectives

## TUKUMBI LUMUMBA-KASANGO

Relationships between society and nature are symbiotic. Our social equilibrium depends on how we perceive nature – the physical environment – and how we use that environment to satisfy our needs. The world is currently experiencing a significant disequilibrium that should be examined critically and understood. The manifestations of this disequilibrium include major wars, massive displacement of people, increasing economic disparity between industrial and developing countries and also between the rich and poor social classes, overcrowded cities, lack of proper sanitation in cities and villages, vicious power struggles, natural disasters, global warming, land pressures, expanding desertification, confusion about major social values, and so on. More than one billion people have seen their environment rapidly deteriorate and their lives become desperate.

There are some estimates that "there may be as many as 25 million environmental refugees in the world today – people who can no longer gain a secure livelihood in their former homelands because of degradation of their soil, air, water and fuel reserves" (Carol Bellamy in "Women and the Environment," The United Nations Environment Programme Magazine for Sustainable Development, Volume 7 number 5, 1995).

At the end of the Cold War, the rise of real and/or apparent democratic movements in most parts of the world has also been seriously challenged by the political philosophy of nationalism, racism, fascism and sexism. Political instability, which also has its roots in the global system, is another symptom of this disequilibrium. One of the main objectives of this conference is to understand the nature of these symbiotic relations sociologically, economically, culturally and politically. Understanding is a dynamic concept. We understand in order to make a planned and positive change. The political, economic

*Tukumbi Lumumba-Kasango is professor of political science at Wells College and Cornell University and was a member of the planning committee for the Third International Ecocity Conference.*

and ecological change cannot occur by itself. It has to be planned. And before we can plan, we have to look at the problematique critically.

I would like to raise some questions. What are the principles that we use or ought to use to manage our space? What are our ends? How do they relate to and interact with the forces of our environment? An international conference on the ecocity should not limit its discourses to arguments about the management in our cities, towns and villages only. We should look carefully at the way we consume our goods, the way we use resources, and at who owns those resources. We are looking at causal relationships between us and the environment. So we have to know the causes of certain things.

In terms of perspectives, I am using what is called systems analysis, and at the same time I am comparing two systems: ecology and community. And in addition, I would like to say it is obvious that the only planet on which we humans are involved in changing our lives through complex processes of constructing and disconstructing our environment is the Earth. From my point of view, the Earth is itself life. Therefore this becomes the major principle in deciding what we are going to do in the environment.

Every system has its particular rules and functions. Its capacity to change depends on many factors, including ability to adapt to new environments or totally new mechanisms to survive new internal and external pressures. Systems are composed of different parts or subsystems, and each part or subsystem has a role to play. The relations between parts are essentially deterministic and complementary. Organically, no single part should be considered superior to another. The equilibrium of any system is a function of each part doing its proper job for the good of the whole. The community as a system is not a natural phenomenon, it is a human construct. It has purposes and a history. Because it is a human phenomenon, it can fail. That is to say that it may embody internal contradictions. Thus, the success of any

Karil Daniels

Tukumbi Lumumba-Kasango

system depends much on how its contradictions are critically understood and how well it is managed. But who should manage and control the system? Looking at ecocities and the prospect of ecocities in Africa will shed some light on these issues.

Although this conference is international in terms of participants in attendance as well as issues under discussion, because we are in Africa, in Senegal, in Yoff, it is logically correct to locate our problematique in Africa with an African historical perspective before making generalizations.

Lives in Africa have been shifted by three major phenomena: slavery, colonialism and the third is a new colonialism. Those forces are determinate forces. They create cities, our environment and ourselves.

African people, like other people, produced cities and urban life before the coming of the West. However, in many ways, philosophically, politically and economically, the nature of the urban life that came with colonization is different from the one that existed before. Although it has been historically established that Africa was gradually becoming part of the global changes that were taking place before slavery and colonization, it should be emphasized that slavery and colonialism, the dominant forces in the world system, forced Africa at the point of a gun into the world system in an elaborately planned systematic manner. My objective is not to describe the philosophy and politics of these two phenomena. Rather, I would like to shed light on the structures of the urban life that they created.

After the Berlin Conference of 1884-85, the African continent was divided between several European powers, notably Belgium, France, Germany and Portugal. The Italians and particularly the Spanish, were unsuccessful in their attempts to join the others in colonial enterprises. As a result of colonization, new structures, bureaucracies and social systems were imposed, in most cases by force

and in other cases through some form of negotiation. Those systems were to support the colonial economy both in the metropolis and in the periphery. The dynamics between the metropolis and the colony (periphery) is part of our concern.

To define what I mean by colonialization, it is first of all the system of separation – separation between the colonizer and colonized, physically, mentally. It is based on the ethos and world view that the dominant values should be the European values. That is very important. The city was the center of power for the colonial enterprise. The relations between the city and the rural area were characterized in the forms of exploitation. In general, the rural area was the place where agricultural products were produced for urban markets. The rural area was considered philosophically inferior to the new system of values, and was politically powerless. The urban markets brought to the rural areas cash, western values systems and Christianity. Within the logic of colonialism and imperialism, the city produced an ecosystem that was universal and was thus perceived as being good in itself.

This view is different from, for instance, historian Ibn Khaldun's, who believed that "Countrymen (sic) are morally superior to townsmen.... Townsmen are so immersed in luxury, pleasure, pleasure-seeking, and worldliness, and so accustomed to indulge their desires, that their souls are smeared with vice and stray far from the path of virtue.... Countrymen, though also worldly minded, are forced to content themselves to bare necessities; they do not seek to indulge their desire for luxury and pleasures" (from "An Arab Philosophy of History" in THIRD URBANIZATION by Janet Abu-Lughod and Richard Hayu, Grove Press, 1965).

This view is qualitatively different from the European colonial perception of urban and rural life. Frantz Fanon described the living standard of the settlers/colonists and those of the colonized in these terms: "The settlers' town is a strongly built town, all made of stone and steel. It is a brightly lit town; the streets are covered with asphalt, and the garbage cans swallow all the leavings, unseen, un-known and hardly thought about.  The settler's feet are never visible, except perhaps in the sea.  But there you are never close enough to see them.  His feet are protected by strong shoes although the street of his own town are clean and even, with no holes or stones.  The settler's town is a well-fed town; its belly is always full of good things.  The settler's town is a town of white people, foreigners.  The town belonging to the colonized people, or at least the native town, the Negro town, the medina, the reservation is a place of ill fame, peopled by men of evil repute.  They are born there, it matters little where or how; they die there, it matters not where or how.  It is a world without spaciousness; men on top of each other.  Their huts are built on top of the other.  The native town is a hungry town, starved of bread, of meat, of shoes, of coal, of light.  The native town is a crouching village, a town on its knees, a town wallowing in the mire."

Bill Mastin

Portuguese colonial structures merge with today's market at Kaolak, Senegal.

The introduction of the international division of labor separated and dislocated the people.  This division of labor is directly a reflection of division of materialism and of mental disposition.  It is also a reflection of power and a system of control.  The separation of town and country can also be understood as the separation of capital and landed property, as the beginning of the existence and development of capital independent of landed property – the beginning of property having its basis mainly in labor and exchange.

Finally, it should also be noted that the extensive division of labor also produced the separation of produc-tion and thus of consumption.  Historically, the appro-priation of capital on a large scale in the city or town in medieval Europe, as well as in the Renaissance and modern periods, created the basis for the acceleration of social tensions as were reflected in the colonial condi-tions.  African cities as well as African economies are strongly peripheral to the metropolitan cities and the world system.  One cannot understand these cities without understanding the world system.

The city is a very complex community inhabited by all.  In developing countries, the majority of people are poor.  In Africa, for instance, there are 620 million people out of which 300 million are poor.  Of these, 200 million are considered extremely poor, according to UNESCO, the World Bank and African governments.  So we have to understand that poverty is a very important element in our economy.  And we have to understand the cause of that poverty.

Debt is another issue that has characterized Africa as a complex community.  As of 1994, the African debt had reached more than 290 billion dollars.  Most countries spent more than 30 percent of their GNP or the revenues from their exports to pay for loan services alone.  If paid regularly, most African states will take 30 to 40 years of all their people's work to be able to pay back those loans.

Now that I have brought the issues to our attention, I would like to propose some guiding principles for creating an equilibrium between our community and our

Bill Mastin

The "door of no return" at Gorée Island, Senegal, through which hundreds of thousands of men, women and children passed onto waiting ships into slavery and away from Mother Africa forever.

physical environments. They include: redesigning the world economy that should be environmentally safe and productive, and also locally self-sustained; recognizing the limits of nature; protecting oceans from too much fishing and creating new policies and practices for the health of fisheries; reducing our passion for consumerism; searching for harmony between the ecosystems and our built habitat; revising the dogmas of property rights; and educating the people about the dynamic relations between humans and nature. We have to collectively work to reduce the level of social and economic inequality that has characterized the dogma and the functioning of the world economy.

As Sandra Postel said ("Carrying Capacity: Earth's Bottom Line" in STATE OF THE WORLD REPORT, 1994), "This chasm of inequality is a major cause of environmental decline; it fosters overconsumption at the top of the income ladder and persistent poverty at the bottom. By now ample evidence shows that people at either end of the income spectrum are far more likely than those in the middle to damage the earth's ecological health – the rich because of their high consumption of energy, and the poor because they must often cut trees, grow crops, or graze cattle in ways harmful to the earth merely to survive from one day to the next."

Redesigning the whole economy at once is a utopian and a logically and historically difficult proposition. However, the reorganization of the local economy within the framework of local possibilities and needs – for instance, creating permanent jobs for people at the local and regional levels and not separating the ethos of the workplace from that of living and social places – should be encouraged. Local initiatives and knowledge should be critically examined and revitalized.

In addition to the above elements, it should be stated that people may respect the values of their environment if, first of all, they know what those values are, and second of all, if they can control the destiny of their environment. That is to say, people's participation in the decision-making process and in the implementation of policies concerning the environment is a sine-qua-non for envisioning stable relations between our built habitat and the environment. This way of thinking is not new in many traditional African settings. However, political democracy without economic democracy will not help solve the major problem that we face today in Africa, which is poverty. And in the North we also have poverty, in the form of homelessness, for instance. This is the parallel that will tell us where to go from here.

This democracy should not limit itself to electoral processes. People should have the power to redefine their values in relation to the nature of their environment and the dynamic of the global and local economy. Sustainable development is development with a human face, or development in which people fully participate. This kind of participation has to be redefined.

In short, we should start thinking about the possibility of creating an economy in which natural, social and cultural resource bases can be preserved. Our life depends on exhaustible biodiversity in forests and mountains, and on resources from the seas and energy from the sun and wind. Thus, it is imperative that we create an economy in which we can develop new relations with the natural forces; an economy in which we can protect the environment, use raw materials wisely and efficiently, convert wastes into resources, integrate needs into the processes of production and build coherent relations between needs and social values, as suggested by the vision of "think internationally and act locally."

# The Ecocity Movement – Deep History, Moment of Opportunity

## RICHARD REGISTER

From the very first dawning of human imagination, the ecocity was there. The animals have homes, all living creatures have homes. Some build their homes: bird nests, beaver dams, bee hives.... Many of the more sociable animals build their homes together. We people are very social animals.

And so, the ecocity impulse gave birth to the desire to build something larger than any of us. It would be a home at home in nature. From the beginnings lost deep before history, we humans could never understand why we couldn't have both nature and our culture at the same time. It wasn't greedy, it was only natural.

Historically the village gave birth to the city. The small one provided shoulders for the large, the old for the young. And now the city threatens to ruin us all: village, town, city, country, plants, animals, forests, seas, climate, ozone layer, resources, us, all. Everything is threatened, not only through massive overconsumption and over-population but by the spreading, sprawling city of cars, highways and massive energy use.

But this is not the fault of the city, though we see Dakar marching across the sand to surround the traditional villages like an army. It is the fault of a particular *kind* of city which we can see all too clearly all around us. But we can build another kind of city, one that enriches the soil and builds biodiversity. We can build such a city if we think clearly about it, dedicate ourselves to the cheerful task and begin.

The present phenomenon we experience as a movement coalescing is a recent set of changes growing from a history that looks something like this:

The early pedestrian village gave rise to the pedestrian city at Çatal Höyük in present day Turkey almost 9,000 years ago and possibly at a few other as-yet-undiscovered sites. The vehicle entered town with the first animal-powered carts and wagons. By the early 1900s the streetcar suburb began the serious outward expansion of cities, began the creation of communities economically disconnected from their own land and dependent upon commuter transportation and the organizing and

producing powers and genius of the nearby city. Meantime, the industrial city with its cramped quarters, severely exploited workers and throat-burning pollution lead to the escape from the city to the ever more distant suburbs.

Near the beginning of the 20th century Henry Ford, more than anyone else, redesigned the city; not Frank Lloyd Wright, not Le Corbusier, not the Garden City designers of England. When Ford invented the assembly line that made automobile mass production possible, he began the now almost universal transformation of cities, creating a new anatomy of scattered areas of single-use development, cars, highways and oil technology. This process is accelerating still, as developing nations around the world, notably China with its soon to be potential one billion drivers, are preparing for their revolution in road building and automobile manufacturing.

After the Second World War, suburbanization in America accelerated rapidly as 11 million men in arms were re-integrated into the economy building suburban housing, cars and the "National Defense Highway

Richard Register

Sprawl, Albuquerque, New Mexico.

System," that is, freeways – and, after barely a deep breath, the new military of the Cold War. In 1950 approximately half of all the cars in the world roved the highways of the United States, the country that then boasted a mere 5% of the world's population. In the 1970s, due partially to the more energy efficient cars produced then, the "Second Suburbanization of the United States" moved not only the people and most shopping to the frayed out distant fringes, but commerce, business and industry as well – right on top of some of the world's best farmland. To get anywhere on this scattered landscape, to work, friends, shopping, even scraps of nature, long trips in cars or longer on transit became a rigid necessity. The result has been enormously damaging for ecology and society alike.

Overpopulation is seen as a major world problem by many and so too for over-consumption, and so too for dependence upon polluting, resource depleting technologies. Biologists and population writers Paul and Ann Ehrlich summarize this much with their formula: $I = PAT$, (Impact on nature, society and resources equals Population multiplied by Affluence multiplied by (the effects of specific) Technology). But we should understand by now that the form and function of the built community itself, from village to city scale, is another of the Big Four Dimensions of the Problem we face on behalf of Earth's life systems. If "Land-use/infrastructure" or "Landustructure," as I call it, is the basic anatomy of the city, then the revised formula should be $I = PLAT$, with the inclusion of the built community, which is, after all, the home of the population, the engine of affluence and the shelter and physical organizing framework for most of the technologies.

We have been slowly awakening to this growing disaster in city building. Lewis Mumford was among the earliest to sound the alarm against cars and sprawl. Ian McHarg gave us mapping tools for better city and regional planning in his book DESIGN WITH NATURE. Paolo Soleri spoke of the place of cities in evolution and took these ideas to their logical conclusion, proposing tall cities of small land area, like space age, single structure variants on American Indian pueblos or North African kasbah towns. His summation of the modern problem as one of building the flat city when we should be building a much more three-dimensional form is one of the clearest formulas for getting started with thinking about ecocities yet identified. In fact, always one to build what he preaches, an artist as well as a philosopher, he began the construction of his experimental town, Arcosanti, Arizona in July of 1970. Three years later E.F. Schumacher warned us about our fixation on the large scale and high powered in his book SMALL IS BEAUTIFUL and reminded us of the virtues of small acts in service to ancient and future values in balance with nature. The flourishing of the environmental movement after Earth Day, 1970 and the "appropriate technology" movement with its emphasis on renewable solar and wind energy, bicycling and recycling, organic farming and cooperative distribution, provided the opening to the subject of re-design of towns and cities to all those concerned about the environment.

Throughout this period, the vision behind spreading suburbia was, ironically, the ecocity impulse, the desire to have nature and city at the same time, and the intuition that it would be healthy. By moving to the suburbs we would reunite with nature while taking with us transportable and broadcastable vestiges of the community in town, vestiges delivered by car and gasoline, and radio and television. The trouble was that vestiges were a pale substitute for the blood and breath of the real thing.

By 1990 a number of us, including myself, believed that the time had come to transform the ecocity impulse into the ecocity insight, that is, to let people know that the only way to have our city and country too was by way of redesigning and rebuilding the city, town and village. We decided to attempt to consolidate the diverse strands, and if possible, forge a real ecocity movement. To do this we organized the First International Ecocity Conference, which assembled many of the pioneers in theory and action. That conference led directly to the Second

Ivan Pintar, Cosanti Foundation

Arcosanti, Arizona, designed by architect Paolo Soleri, under construction since 1970, is the first full-fledged ecological city project. Standing midway between the Indian pueblo past and the ecocity future, it provides an existing and growing foundation on which to build.

International Ecocity conference in Australia, then to the Third here in Yoff and Dakar. In addition to those of us concerned directly with rebuilding communities large and small, others have focused on the architectural response to environmental problems. The European Eco Logical Architecture Congress, held in 1992 in Stockholm and Helsinki, for example, and the INTERARCH Conference in Sofia, Bulgaria in 1994. The United Nations Conference on Environment and Development held in Rio de Janeiro in 1992 barely dealt with architecture or town planning at all. But the EcoVillage Conference at Findhorn, Scotland in October of 1995 opened up the rebuilding of villages on behalf of our whole civilization, with small emphasis on cities.

Today we have many assets for our movement that can be used and supported in various ways. The material assets can be expanded, such as the tools and the actual projects – many of these you will hear about at this conference – from the establishment of new urban rail lines that Jeff Kenworthy is scheduled to tell us about to the installation of the separating toilets of Anders Nyquist's villages designed around ecocycles.

The "human assets" – those people doing the dedicated, creative work – can be supported emotionally, with material assistance, and financially. They include people walking and biking and pointedly *not* driving and people creating rooftop gardens; solar greenhouses; creek, river, shoreline and wet lands restoration projects; urban community gardens and street orchards; street traffic calming projects; new streetcar lines; and urban greenbelt projects from Stockholm's on. They include people writing the "Green Plans" of the Netherlands, and New Zealand; artists like Friedensreich Hundertwasser who built Hundertwasser Haus in Vienna, Austria and designers, builders and residents of cohousing projects around the world, notably in Denmark where they began, and in the United States where they are replicating quickly. The people building the Halifax Project in Adelaide, Australia, and the ecovillage neighborhood in Los Angeles – an effort of the Los Angeles Ecocity Council and Lois Arkin – are international treasures, and so are those building the rural ecovillages, Curitiba, Brazil – perhaps the world's best example of a city attempting to transform itself into an ecocity – and Arcosanti, the good old first thoroughgoing, no-holds-barred experiment in ecocity building.

The old traditions offer numerous assets too, from the Garden Cities of the turn of the 19th to 20th century, back through the pedestrian villages of the world like Yoff, Ngor and Oukam, the kasbahs, and Venice, Italy, to the Indian Pueblos and the earliest urban archeological finds in Turkey. Finally, among the most valuable assets that the ecocity movement possesses is... *us.*

Now we are up to date in a very general way. And now we are beginning the future by sharing our ideas, trading some new and some tried and tested old tools and re-committing ourselves to following through in action. The organizers of this conference and we who have supported them in this enterprise hope that when the curtain goes down on this event, the theater doors open to a whole new and very effective network of rejuvenated friends and colleagues, well equipped and ready to build the city, town and village of the future.

Richard Register

Depaving! Beginning the roll-back from sprawl. This depaving project made way for a small garden for University Avenue Homes, a housing project for the recently homeless in Berkeley, California.

A Curitiba park. When money was available from the federal government, Curitiba bought land along the rivers rather than culverts in which to bury them. Floods just rise and settle harmlessly in the parks.

*Curitiba, Brazil has initiated probably the most comprehensive ecological city planning and policy development of any city in the world. From innovative public transit, to recycling, from ecological education in the schools to creation of pedestrian street environments, from broadly accessible planning forums to tangible and timely follow-through in action, Curitiba radiates imagination and commitment. It is as complete a model of ecocity beginnings as we have to date.*

# 2. Toward the Ecocity – Cities in Transformation

City governments are pivotal in the process of rebuilding cities upon ecological principles. Some of them are embracing the Agenda 21 agreement of the United Nations Conference on Environment and Development in Rio de Janeiro in 1992, and now they commonly use the term "sustainability" in describing their own intentions. They represent early strides toward a new integration of culture and nature, economics and ecology, and what our three speakers representing three such cities have to say is crucial information. And so is information from our one speaker presenting major land use and transportation changes in cities around the whole world. It's crucial because it represents those empowered officially and empowered by their citizens to begin building a new kind of city, the ecocity.

You will see here that ecological restructuring has in fact begun. There may still be a powerful drift toward extension of the auto/sprawl/highway/cheap energy city of environmental poisons and fractured community. But a powerful – and the point of this chapter – an *empowered* movement has already begun a transformation in the opposite direction, toward, as Jeff Kenworthy calls it, "the walking city."

Prof. Kenworthy, from his university post on the lonely western fringe of Australia, studied cities around the world. He looked at their most basic structural foundations in their land-use and transportation patterns. Then, to spread the warning of crisis – and the good word of those making progress against it – he launched out around the world to speak and photograph and gather facts, and speak and photograph and gather *more* facts. Now, he reports in this chapter on dozens of cities making solid progress toward the ecocity, as did his research and writing partner Peter Newman at the First and Second International Ecocity Conferences, in 1990 and 1992.

Perhaps the best known city for its already existing and quickly evolving ecological city changes is Curitiba, Brazil, represented at Ecocity 3 by Cleon Ricardo dos Santos. In his talk and paper we see "on the ground" the widest range of ecological projects achieved to date by any city known to the editors of this book. But Bergen, Norway might catch up. Bente Florelius, a member of the city's planning department, gives us an inside view of their plans for major transformation in Bergen by the year 2020. Josep Puig i Boix of the stateless nation of Catalonya in Spain, and Councillor of Sustainable City of Barcelona, is a City Council Member there and reports on the politics behind their first policy moves toward the ecocity, revealing along the way some of the process and difficult decisions encountered.

With these reports we will have looked under the hood of the machine, to take a metaphor from the opposition, and zipped around for a short test ride. The critics agree. The early model ecocity, as they say in the newspapers' classified car ads, "Runs good; needs work."

# City Building and Transportation Around the World

## JEFF KENWORTHY

Cities started as walking cities. The traditional village everywhere is a walking city. Yoff is a walking city. Ngor is a walking city or village. These places are very high density, very organic in their structure. In these old mountain towns of Europe (*Jeff begins a slide show*) there is virtually no use of motorized transport, none at all because it didn't exist when the towns were built. In walking cities you don't need an automobile to get around, you don't even need public transport because the city is so dense, so mixed in its land uses, so compact, it is possible to walk anywhere. I would imagine that in Yoff and Ngor 80% or 90% of all trips are on foot. We see the historic walking city still in existence in the centers of all European cities. We also see the walking cities still in Asia in cities like Indonesia's Surabaya, in its neighborhoods called the Kampons, the traditional form of Indonesian urban settlement. Within the Kampons the traditional life is based around traffic-free environments where the movement space is also the market space.

The traditional walking city in the West, during the Industrial Revolution, wasn't a very pretty place. But for all the sooty physical degradation it was still a wonderful place socially, with a sense of community and neighbor-

Jeff Kenworthy and Cleon Ricardo dos Santos at Ecocity 3.

*Jeff Kenworthy is a co-author of WINNING BACK THE CITIES and Lecturer in Urban Environments, Institute for Science and Technology Policy, Murdoch University, Australia.*

hood. Unfortunately, this social interaction is something that the West has lost. It is something that Yoff, Oukam and Ngor have to teach the West: to rediscover community, belonging, and neighborhood.

When public transport came along, first as the horse-drawn streetcar or tram, then the steam train, then the electric train and electric tram, it caused the city to spread out. On our map we can see the walking city at the center with small suburbs located like separated beads on a string, on a few rail lines leading out into the country. The streetcar suburbs grew up in a relatively narrow ring on the fringe of the old walking center. Both electric trains and steam trains served the stations where the nodes of suburban settlement appeared.

This kind of city with its pedestrian core, streetcar fringe and modest number of rail station-centered suburbs, is not dependent upon the automobile. Land use is inextricably linked to the transport system – and it works ecologically and socially.

You see this in effect in the new developments around Stockholm built since the 1950s. In this slide we see Shista on the electric train line near Stockholm showing how development is clustered around the railway station with much green space all around. The traditional streetcar suburbs of San Francisco still exist and are much less dependent on the automobile than the surrounding suburban sprawl. And in Stockholm again we can see the linear form which the train-based city developed. Everything had to be within walking distance of the tram lines. If you develop in nodes around railroad stations and if you develop in transit oriented forms you have all this green space available. You do not destroy your urban environment with suburban sprawl .

Unfortunately, as Richard said in his brief ecocity history outline, we now have suburban sprawl in all our cities: the automobile city. And the automobile city destroys the land use transport connection. It makes development footloose, development can go anywhere it wants. We have to pump huge amounts of resources and energy into running these kinds of cities.

Shaffhausen, Switzerland, a traditional compact "walking city" common in European city centers today.

So we may still have remnants of the old walking city at the center with transit suburbs in many cities, but the transport systems are dominated by the automobile. And this is what we end up with – Los Angeles, a total disaster, virtually no community, smog covering the city just about every day. A very problematic urban environment.

My writing partner, Peter Newman, and I have spent some 17 years analyzing transport and land use patterns around the world to see how dependent cities are on the automobile. Briefly summarizing our work, we studied 32 cities, we looked at U.S. cities, one Canadian city, Australian cities, European cities and Asian cities.

What we labeled "class one" cities on the charts have very high automobile dependence with practically no role for public transport or cycling or walking. They all have very high energy use. These are the Phoenixes, the Houstons, the Denvers, the Detroits of the world. Perth, the town I live in in Australia, is in this category. Moving across the first chart from left to right, car dependence gets less and less. The cities in class two tend to have very good rail systems. Toronto and New York are the transit capitals of North America . Next to the right in the third and fourth categories you have most European cities. They have low automobile dependence, with public transport about equal with cars. In the last category at the far right edge of the chart, with very low automobile dependence relative to transit, we have Tokyo, Singapore, Hong Kong and a couple of European cities.

If you look at the factors that determine these patterns you find that density is a very important one. In U.S. cities and Australian cities, urban densities are very low, about 13 or 14 people per hectare or 5 or 6 per acre. Here low density development eats away everything like a cancer. It is almost impossible to stop. Canadian cities are about twice that density. European cities are higher again and the Asian cities, both the wealthy ones such as Tokyo, Singapore and Hong Kong and the poorer ones such as Surabaya, Bangkok and Jakarta are still the same density and you can see at these high densities what a difference it makes to transport patterns.

How this effects car use is immediately obvious in our third chart. If we look at American car use in 1990, we find that it is absolutely massive. About 11,200 kilometers per person per year compared to the developing Asian cities like Surabaya and Jakarta with only about one tenth that level of car use. The really worrying thing about this graph is the rate at which American cities are growing in per capita car use. In ten years they have increased by nearly 2,200 kilometers per person which is equal to the total car use by the average person in Tokyo.

What about transit use? In U.S. cities about 3% of all passenger movement is on public transport and it has declined from 3.7% from l983 to 1990. Australian cities do a little bit better. Toronto does significantly well with 25% of all travel on public transport. European cities also do reasonably well. But the wealthy Asian cities with

Richard Register

Dave Cohen of Ped Ex, the pedal-powered delivery and courier service of Berkeley, California.

levels of wealth equal to the West have 65% of travel on public transport. The developing Asian cities are a big worry because they are adopting the automobile at a great pace. Only 40% of travel is on public transport.

If we look at density in relation to per capita use of cars, we find a systematic change. As the density increases, the use of public transport increases and the use of cars decreases. So density is a very important factor in understanding how villages like Yoff work in terms of their walking and social life.

Looking at energy use we see that gasoline consumption per person also varies dramatically with density. Tokyo, Hong Kong, Singapore have very low per capita levels of gasoline consumption. And of course, American cities are on the top of the graph.

Our fifth chart, about walking and cycling to work, shows that only 5% of the people in U.S. and Australian cities walk and cycle to work. Nearly the same in Toronto. But very much higher numbers of people walk or cycle to work in European cities and higher again in both the wealthy and developing Asian cities. As the city becomes denser the amount of walking and cycling increases.

So if we want walking and cycling we have to create the environment that is amenable to walking and cycling. We have to get the car under control. And they got the car under control in the Kampons of Surabaya. The streets there are so narrow that cars can't squeeze in at all. The Kampons function not only as walking and cycling environments, but with fruit trees in the streets, as food producing environments too.

What about roads per person? In another chart we see that the car-based cities provide huge amounts of road infrastructure. Canadian cities have about half what

we provide in Australia, the European cities lower again, and the wealthy Asian cities lower again.

Chart number seven is about parking, another Western disease. Parking covers huge areas of our cities. U.S. and Australian cities provide enormous amounts of parking per thousands of jobs in the central business district. Tokyo only provides about 33 parking spaces per 1,000 jobs. Everybody gets there on foot or bicycle or public transport. But the developing Asian cities are a worry because they are very high in this respect.

So if we want a future for our cities like this tangled mass of freeways and sterile buildings in this picture of Detroit, Michigan, then go ahead and build lots of central parking. Such a wonderful place to be! If you want good environments then pedestrianize your city centers, get the car under control.

I think we are all pretty much aware of the problems of automobile dependence: smog, urban sprawl, vulnerability to oil supply cut off and price increases, storm water problems, traffic problems. We have a whole series of economic problems related to the loss of productive rural land, the loss of urban land to bitumen and concrete, and a whole series of social problems: loss of street life, loss of community, loss of public safety, isolation, people being cut off from one another, crime increasing because there are no natural checks on the urban community. And we should not forget that we are prepared to go to war over oil – and have. This is a real threat as we enter in the next century. Nations are becoming more dependent upon oil from the middle east.

In this photo of Salzburg, Austria, we see a scale drawing of a typical freeway interchange superimposed. It's enough to wipe out the entire old pedestrian city of Salzburg. Automobiles destroy neighborhoods, destroy contact between people and sever neighborhoods. Bangkok, a disaster, the Los Angeles of the east – it presents an image for the future of our cities, of developing country cities in particular, that we should avoid at all costs. Anywhere you go in Bangkok you are assaulted by air pollution. You cannot walk along the street without holding a handkerchief to your mouth because of the suspended particulate pollution.

How do we stop this? I want to discuss three things: developing better transport systems, traffic calming, which is reclaiming space from the car in favor of other modes, and especially I want to talk about urban villages, which do exist in modern forms all around the world.

Light rail is gaining favor in cities around the world because it is cost effective and flexible. In Hannover and Frankfurt, Germany it can operate on the street and turn the corner almost at right angles. It is also very amenable for people in wheel chairs. You can get into these vehicles; they are low-floor vehicles. Light rail systems can carry bicycles as well. We have very people-friendly urban environments that are built around light rail systems. In Freiburg, Germany the rails are set like two thin ribbons in a grassy lawn. In Portland, Oregon we have widened footpaths and auto traffic reduced to only one lane with the light rail system very close.

So what about traffic calming? What about sharing streets between pedestrians and other uses? Here in Frankfurt we see a traffic calmed street. If you can't get the car out entirely you can at least calm the traffic. The Bachenheim district in Frankfurt is being totally traffic calmed. Main roads can also be traffic calmed. One of them in central Frankfurt still carries traffic but has good facilities for pedestrians and cyclists. The Kampons of Surabaya, pictured here, are a pleasantly extreme example: top speed 5 kilometers per hour or about a leisurely walk. The speed limit is controlled by the traditional community. The street is used for markets, for non-motorized transport and food production, a complex mix of land uses in a mostly residential environment.

Now let's look at recent urban village development. This is what we are trying to do in the West. In Munich a new urban village is built on a subway line, traffic free. There are gardens on rooftops. No cars here. The cars are underground. The streets are for human life, places for children. You know the children in Yoff and the villages around Dakar are wonderful – to have them in the streets, to be able to play in their own neighborhood is a wonderful thing; in a number of places we are trying to bring this back in Western development. Freiburg has another example of an urban village called Seepark, built around a large lake and a light rail line with a grass track bed. There is no incursion of traffic into this community. It is totally traffic free. All movement is on foot and bicycle. Children are safe in these kinds of communities. With high density there can be lots of open space.

Now let's look at an urban village in Zurich. You can see its red roofs, the light rail line coming through, an electric trolley bus going in the other direction. The community is built on ecological principles. Storm water

Jeff Kenworthy

A fruit tree-lined street in a "kampon" of Surabaya, Indonesia.

is handled on site, the buildings are designed around passive solar principles, very well insulated and so on.

Here we have people walking from that development to the public transport system. If you pull back from that urban village you can see how much green space is around Zurich. It is a wonderful city for the preservation of nature and for community gardens. If you have this kind of development, not sprawl development, you have the possibility to preserve the green space in a communal way. And you have forests right next to the houses, beautiful forests. No part of Zurich is more than ten minutes away from a forest that surrounds the city.

For a North American example, we go to the city of Vancouver in British Columbia. The False Creek development is an urban village of grand proportions. Ten thousand people live in this urban village, and it is a totally traffic free community. This is not an architectural plan, this is not a drawing, this is reality. This is happening and this is the wave of the future for urban development. Good environments which allow people to mix with one another, to experience walking and cycling and to be safe, especially for children to be safe, to be set on water without the impact of noise and fumes.

Vancouver is very good at integrating village developments into their train systems. You can see the train through the arch beyond the playground equipment, right beside the trees. For local needs here you walk and cycle. To go out to dinner and a movie, or to a friend on the far side of town you get on the train. You do not need to own an automobile to live in this community and it is a beautiful community with a lot of care for the public space, for the common good, for the common heritage. People are able to meet in public places because there is no traffic, there are no car fumes, no mechanical noise.

Effective public planning creates these developments. They are not just happening through market forces. There is very strong public planning, which is making these developments a reality.

So how might we restructure the automobile city? We need to invest more in public transport. We need to extend the railway system which provides the focal points of the clustering of development. We need to build urban villages and within these villages, we need to ensure that it is a walking and cycling environment, not an environment for the automobile.

I have a series of indicators which show the directions cities need to take, specific annual objectives which cities could use to gauge their performance in land use terms, in terms of public transport objectives. Each year we can assess whether or not we are going in the right direction. These are quantifiable measurable parameters we can use. Are we building more pedestrian streets? Are we getting the car under control? Are we building less infrastructure for cars? Are we going down in our car use or are we going up?

The Skytrain elevated rail in Vancouver, Canada, connecting downtown with the False Creek development.

# The City as a Whole Organism

## CLEON RICARDO DOS SANTOS

Cleon Ricardo dos Santos

All projections of population growth on our planet indicate that in the early 21st century the major part of Earth's population will be living in cities. Yet despite the simple fact that cities concentrate huge numbers of people in small areas, the "built environment" does not seem to be of great concern to national governments, nor does it appear to be a problem in most people's minds.

But in fact it is. The Environmental Advisory Council of Sweden has studied the built environment from an ecological perspective and recently reported that "cities are the major contributors to the degradation process caused by human beings, and they are growing frighteningly." Pondering the initial statement, the report elaborates saying, "many of the environmental issues related to sustainable cities concern infrastructure. Supplying cities with water, food, energy and transport creates several environmental problems, including garbage and air, water and noise pollution. To solve environmental problems we must adapt urban supply systems to the environment."

I would add that it is imperative to educate citizens about urban environmental concepts and simultaneously provide them with guidelines for participating in the building of more environmentally healthy cities. In this regard, the experience Curitiba has developed over the past twenty years may be worthy of some attention.

These are some of our greatest successes:

– The city established a bus system that is so advanced it nearly matches the people-moving efficiencies of metro subway systems, but at an infinitely lesser cost.

– Instead of burying its five modest sized rivers in concrete pipes, – the usual practice at the time – Curitiba purchased stream-side land in the 1960s with federal flood control money and built a system of parks that

magnified the city's green space many times over. This also allowed storm water to swell harmlessly into the lower parts of the parks, putting an end to the floods that periodically afflicted the downtown.

– Curitiba established a garbage pre-separation program that allows for one of the world's most efficient recycling operations for paper, glass and metals while greatly extending the life of the municipal landfill.

– And the city created an institution to disseminate information on environmental issues to the population at large – the Open University for the Environment – while expanding environmental education programs geared to children from all social strata.

I will survey this 20 year history next, but first briefly glance even farther back in history. The city of Curitiba was founded in 1693 by Portuguese gold hunters, and experienced a rather slow growth until the end of the 19th century. During the early 20th century, a significant increase in population took place, leading the city to devise its first master plan in 1942, under the guidance of French urbanist Alfred Agache.

During the 1960s, pressed by the rural exodus, which in turn was caused mostly by the mechanization of farming in the State, the city underwent an accelerated population growth, jumping from 360,000 to 600,000 inhabitants, leading to the need to revise its planning. By 1995 the population was over 2 million.

The Preliminary Urban Plan, prepared in 1965, set forth the three main lines of action: first, linear city growth along five structural axes, second, revitalization of the inner city, and third, priority to humans, not to automobiles. An urban transformation process based on these premises, unprecedented in Brazil, was soon launched with the support of the Institute for Urban Planning and Research of Curitiba.

The first step was to close the main street to traffic and return it to pedestrians. Automobiles were replaced

*Cleon Ricardo dos Santos is General Coordinator of the Open University for the Environment in Curitiba, Brazil. He was and remains part of the inspired team of creative architects, administrators and activists in Curitiba who have brought their city as close to an ecocity as any on Earth today.*

Light pours in under the canopy of trees and onto strollers enjoying a Curitiba park.

with flower pots, newsstands and other urban equipment. Business people worried about lost patronage and car drivers decided to stage a drive-through protest to reclaim the street for themselves. But the office of the new Mayor, Jamie Lerner, summoned city workers who rolled out watercolor paper, invited school children to the street and handed out brushes and paint. The children – and the pedestrian street – emerged victorious. Within days business was up and people were beginning to be proud of their new street.

This action was followed by a reordering of existing streets, which strengthened the five main axes of travel from the outer parts of town to the center. The center lanes of these streets were turned into high speed bus ways, and the new arrangement was accomplished without major expropriations of land or buildings and without expensive engineering. It was easily integrated into the bus system by way of new boarding stations that allowed users to take different routes with a single fare.

The higher speed streets for the "express busses" were kept out of the downtown center, passing just outside of it. Meantime, a network of bicycle paths was created, mostly parallel to the bus routes, and higher density development was encouraged within two blocks of the bus route and discouraged farther away.

With this new transportation system and its associated land use policy, bus speed increased from an average of 5 - 6 km/hr to 18 - 20 km/hr. The integration of the "direct routes" with the previous bus system, the availability of single fares for multiple transfers between lines and the introduction of novel "tube stations" led 27% of car owners to opt for the mass transport system.

The "tube stations" deserve special comment. The designer was none other than Jamie Lerner, Mayor at the time and an architect by training. The stations are like giant glass bottles – about nine feet in diameter – and oriented horizontally on their sides, with two open ends. Steps with a handicap lift option lead up to these open ends and inside is a turnstile and place to insert coins for tickets, and tickets to open the turnstile. In the middle of the glass tube is a long bench, and facing the street, large double doors. Thus people buy their tickets, enter, and sit in the transparent station sheltered from the weather. When the bus arrives, large double doors open to allow rapid entry. No need to juggle coins and tickets with the driver. The saved time, increased convenience and stylish station design has made the bus system very popular, leading not only to reduction in car traffic but also, according to the International Institute for Energy Conservation, to about 25% less fossil fuel consumption in Curitiba than in the average city of similar size.

Most recently, Curitiba has introduced triple "bi-articulated" busses, busses that become almost like short, above surface subway trains on rubber wheels, and operate at similar efficiencies.

Curitiba's environmental programs should also be highlighted. It is worth stressing that the system of parks and gardens established over a 20 year period allowed an increase from one half a square meter of open green space for each person to fifty square meters – even though in the same period the population had jumped from about 600 thousand to 1.3 million. This included several urban parks, among them the Iguaçu Park, one of the largest urban parks in the world, featuring the grassy temporary flood containment basins and complete with English greenhouses and a French formal garden.

Since 1991 the city has passed legislation based on accumulating experience for the purpose of preserving green areas and penalizing indiscriminate tree cutters, defining the activities that need environmental licenses to operate and strictly limiting urban noise.

The very successful "garbage that is not garbage" program must be mentioned. A public campaign launched in 1988 convinced the population of the economic and ecological advantages of re-utilizing and recycling garbage. In poor areas with narrow roads impenetrable by recycling trucks, people were offered packages of food and transit tickets in exchange for recyclable materials. Since the program was set up, the level of childhood diseases detected in the municipal health care centers in these areas decreased significantly.

On the first day of the program the mayor donned a garbage collector's overalls and went to work with the other sanitation workers, saying that if it wasn't beneath the dignity of the mayor to separate and recycle then it shouldn't be beneath the citizens' either. Once the recycling system was in place, around 30 tons of garbage began to be recycled every day. Roughly 70 percent of paper and 60 percent of plastic, metal and glass are recycled in Curitiba. Within three years, the amount of waste going to Curitiba's landfills decreased 30% in weight and 50% in volume.

The program has another feature that has been particularly successful, sometimes referred to as "recycling people," in which the homeless, street alcoholics and others in desperate financial and sometimes emotional shape have been hired at decent wages in the recycling facilities that complement the separation work provided by the people in the neighborhoods.

Finally, it should be mentioned that environmental education has deserved great attention on the part of the municipality of Curitiba. Programs entailing community participation, such as "Garbage that is not Garbage," could not have materialized without the involvement of children who were already educated in environmental concerns in their classes. As you might expect, in many families, the children learn first in this manner, bringing their parents along into the programs after them. A

A school vegetable garden in Curitiba.

troupe of four actors dressed in green outfits visiting one grade school after another, called "The Leaf Family," dramatized the ecology story, convinced the kids, who in turn convinced their parents.

The Open University for the Environment takes the direct educational approach to adults. It is located in a building – in which I have my own office – that is made of recycled telephone poles and set into an old quarry. Once a real scar on the land, this place is now a green paradise with a small lake and its ducks and swans. In the University, students learn technical and design skills for credit and leading to employment, or simply exercise their imaginations and citizenship, gaining knowledge to help Curitiba be a better place. Many opinion leaders meet here to learn about ecology and alternative solutions to environmental problems, and the school also serves as a think tank for new ideas, continuing the tradition for which Curitiba is beginning to become well known.

As a conclusion, we must stress that Curitiba is not an island within the Brazilian reality and that it is also affected by the economic crisis the country has been facing. Nevertheless, its managers believe that with good resource management allied to the continuous creative problem-solving a lot can be accomplished so that cities can live more harmoniously with nature.

# Barcelona's Political Approach to Sustainability

## JOSEP PUIG I BOIX

Barcelona is a city located at the northeast coast of the Iberian Peninsula, in the northwest corner of the Mediterranean Basin. The city is the capital of a nation without a state, Catalonya, with its own culture, history and its own language, Catalan, a different language from Spanish.

The city has 1,614,000 inhabitants (1995). Our population has stabilized in this last decade with a downward tendency. It reached a peak of 1,907,000 inhabitants at the end of the eighties. However, production of municipal solid waste and energy and water consumption have increased.

Around Barcelona there is a Metropolitan Area of 300 square kilometers. All together it represents 3 million people – half of Catalonya's population living on 1.8% of Catalonya's surface.

Those 3 million people produce 1,250,000 tonnes of solid waste a year or 820 pounds each. Half of that is organic waste, which is incinerated or placed in landfills. 30% of the city's contribution to the greenhouse gas warming comes from methane generated in the landfills, 30% from the transport sector. Of the energy used in the transportation sector, automobiles are responsible for 96.7%. Public transportation is responsible for only 3.3% of the transport energy. That includes an underground metro, busses, and taxi cabs. Taxis use about a third as much energy as the metro and busses combined.

Barcelona is very low in $CO_2$ production. For some comparisons, Copenhagen produces 2.5 times as much $CO_2$ per person per year, Hannover 3 times as much, Toronto 5 times as much and Denver 7 times as much. However, in addition to policies that are striving for

*Josep Puig i Boix is Sustainable City Councillor for the Municipality of Barcelona, Catalonya, Spain. He is the reluctant politician who never thought of holding public office until his friends insisted he run for City Council to represent the kinds of ideas assembled at Yoff. He won. This contribution is a compilation of his talk, one of his papers and some information he provided in correspondence.*

improvement, there are some special reasons for these statistics: most heavy industry has been transferred out of Barcelona in the last decades, the Mediterranean Sea helps save energy because it maintains a relatively pleasant climate, saving energy for heating and cooling, and 80% of the electric energy produced for Barcelona is from nuclear power plants! (*Mr. Puig i Boix's exclamation point!*)

Food used to be produced in the immediate vicinity but is now brought from distant places. Barcelona doesn't capture rain water now, but it could. With the present pattern of using water and with the inefficient devices now in use, the amount of rainfall in an average year equals 63% of the water the city uses. But if the consumers used water-efficient devices, the city could meet its water needs with the amount of rainfall that the city receives in an average year. The problem is that it will be necessary to design and develop water-catching systems adapted to a modern city like Barcelona. We are seeking ideas and experiences on urban water catchments and cisterns right now. Please send them our way!

With regard to solar energy, there are no official plans to apply photovoltaic (PV) systems in the city, but many non-governmental organizations (NGOs) and small companies are lobbying to promote PV systems. The present city government has, however, taken steps to install solar thermal applications in domestic and public buildings. All the buildings planned and built by the municipal body as public assisted housing are being bioclimaticly designed and are incorporating solar water systems (though only a small fraction of the houses on the market are built by the city).

Financing is the big problem for solar energy and renewables. My proposal, supported by the Civic Table on Energy – more on that commission presently – is to finance these with the same amount of money the municipality is collecting from the electric and gas utilities, which are privately owned. These utilities are paying the municipality 1.5% of all their revenues from the city consumers using electricity and gas. The same

Josep Puig i Boix (second from left) meets with conferees and Yoff planners to discuss plans for Yoff's future extension.

amount of money must be devoted to promote renewables in the city. This proposal is in a process of discussion. I'm building a lobby of NGOs and renewable energy companies in order to support this proposal and get it adopted by the City Council.

In Barcelona, steps toward ecocity solutions evolved politically and ecologically at the same time. From 1991 to 1995 the government of the city was composed of two parties: the Catalan Socialist (PSC) and the Catalan New Left (IC). During this time and backed by the mandate of these two parties, and because of the pressure of a coalition of local NGOs, the city of Barcelona adopted its First Environmental Program consisting of 20 programs with 75 actions. One important result was the creation of a new municipal body, the Civic Table on Energy, or "Barcelona Estalvia Energia" (Barcelona Saves Energy), where representatives of civil servants of the city and representatives of the NGO platform meet.

Elections in May of 1995 produced a new government by July of that year. Two new parties entered in the local government: the Catalonya Greens (EV) and the Catalan Indepentist Party (ERC). The city reorganized to have 10 plenary commissions, each one with a president and a vice-president. In order to keep political responsibilities and technical work separated, all the technical services were reorganized too. At the end, the result was ten political commissions and five operational, or technical, sectors. In this new government, a new responsibility was created: a Councillor of Sustainable City, a position that I presently hold.

In November, 1995, the City Council unanimously adopted The Charter of European Cities and Towns Towards Sustainability (The Aalborg Charter).

But what has happened with the First Environmental Program? One of my first actions as the Councillor of Sustainable City was to ask technical services to report on the Program. The surprise was that from 75 actions listed in the Program, only 27 were accomplished and 12 partially accomplished. 36 weren't put into practice at all. It will be very interesting to learn in detail why this happened. Only then will we know the reasons for both action and inaction, and then be able to elaborate the Local Agenda 21.

One of the reasons could be the lack of continuous exchange of information between local NGOs and the City Council. I have been working during the last six months in this direction, both stimulating The Civic Table on Energy created by the previous government, and making contacts with as many local NGOs as possible. All the contacts were successful and by now we are working together to create a commission of civic participation in sustainability so that all sectors of civil society can reach a consensus.

What are the plans for the near future? According to Aalborg Charter, it is necessary to create a special working group (called The Environmental Forum) to involve as many interest groups as possible. This Environmental Forum will be the commission of civic participation in sustainability that I will be working with as Councillor of Sustainable City.

But we must face an important problem, because there are two positions about how to proceed: Position one: the government will draft a scheme of environmental problems and will ask a group of experts to draft a document. This document will be submitted for discussion to the Environmental Forum, which would have been constituted by the City Council in the meantime. Position two: The City Council and NGOs together will first create the Environmental Forum and then draft the statement of environmental problems. Then proceed together building a consensus. In the next months a decision will be made.

# Bergen, Norway –
# Plans for the Year 2020

## BENTE FLORELIUS

This conference has assembled participants from different nations and many different backgrounds. And we are gathered here in Senegal, a country that is very different from the ones most of us come from. My own country, Norway, is a rich industrialized country in the far north of Europe. There in our planning department of the city of Bergen we are asking how environmental problems affect us and we are trying to decide what changes we can make to ensure that our country and city continue to develop, but in a positive and sustainable way. Does it mean taking on new roles, and using new working tools and mechanisms? Can we define a "sustainable local community?" Can our vision of the future lead to constructive debate in a social democratic society? I will try to throw some light on these questions by referring to two projects, one for a suburban area transforming into a real community centre, and one for transforming a large area of our downtown.

In a global context, all of us from our different communities have common problems which we must try to solve together – that's why we are here. But we are each at different stages of development, with different climatic conditions and histories. Our societies are also differently organized. For these reasons we must each approach the problems in a different way.

By world standards, Norway is a small country, but in relation to our size we have been able to exert a great influence, especially during the last 30 years, following the discovery of oil and gas under the North Sea. Suddenly we became one of the leading oil-producing nations and as a result, a nation with a high material standard of living. Electrical energy from our mountain rivers also contributes greatly to the country's wealth.

As a result of our long-standing social democratic form of government, we are a nation with relatively small differences between the poor and the rich.

Our country consists of large natural and mountain areas that cannot be cultivated, and the population lives mainly along the coast and has traditionally made its living from fishing and small-scale farming, as well as maritime and trading activities. Today, the population is to a great extent centralised around the ten largest towns. Nevertheless, Norway's regional policy continues to stress the importance of maintaining local communities throughout the length of this long country. In this connection, it is important to mention that most Norwegians are still newcomers as urban dwellers; they have an understanding of nature and the need to preserve it that they acquired from their parents and grandparents. It is perhaps precisely this that has to be seized on if we are to get the rich Norwegians – with their full stomachs and high personal consumption, and with enough energy to heat up bigger and bigger houses, and enough money to buy more and more cars – to understand the seriousness of the situation, show solidarity with the rest of the world and change direction with the aim of achieving sustainable development.

What then, are the problems facing us that must be seen as important signs telling us that we must change direction? Spring tides, wind-thrown trees, mountains of refuse generated by our cities and atmospheric $CO_2$ build-up are all such warnings. Spring tides are the highest of tides, which have been happening more and more frequently every year, such that the danger of damage to our historic buildings is becoming quite significant. Wind-thrown, or up-turned trees after storms prove that the strength and extent of storms and natural catastrophes are increasing, and it is the insurance companies who are taking these signals most seriously of all, as their business is directly affected. Refuse disposal has become a major headache and the biggest challenge of all is the increasing production of $CO_2$ and associated climate change. Dealing with this means reducing energy and

*Bente Florelius is Architect and City Planner for Bergen, Norway, coordinator of their Healthy City Project and administrator of a major inner city revitalization project there. This synopsis is based on both her slide presentation and her written paper.*

City of Bergen Planning Department

Bergen used to welcome
its visitors through its
waterfront, its old "front door."

material consumption implying real change in the social fabric all the way down to individual lifestyles.

In addition, I can show you a graph depicting quality of life that shows there has been a decline in important social qualities in spite of Norway's ever-increasing material wealth. This seems to be an increasing problem throughout the entire Western world. It is reflected in criminality, violence, child abuse, use of drugs, break-up of families, psychological problems and loneliness, which suggests, as some have pointed out, that "in a sustainable society the most important source of energy is the warmth of the human body."

And again, another indicator of a growing problem is the declining state of health in certain categories. It is estimated, for example, that about 40% of the children born in Norway today will suffer from asthma and allergic complaints. There are two main causes: first, poor indoor climate which results from toxic building materials and poor building design, and second, traffic pollution.

Now we turn to the city of Bergen in particular, the second largest city in Norway. Once the capital, it has long been an important European centre of trade and shipping activity. Today Bergen has 220,000 inhabitants, and, in a Norwegian context, the city has rich urban traditions. The building styles and cultural environment are typically European. What is special about Bergen is its compact urban development and its location on the narrow coast between the vast North Sea and the high mountains in the west of Norway. Less special is its continued traffic increase and expansion, since 1950, that is covering an ever greater area, even though the population has not risen significantly. Many are now willing to accept traveling for one hour to get to work. The traffic creates barriers and noise. In addition, pollution hangs over Bergen in cold weather. In winter, people suffering from asthma are advised to stay indoors when nitrogen dioxide in the air is far above the danger limit. Our challenge, then, is to improve the standard of life for a largely stable population, but within the framework of a sustainable development.

## A Vision for the Year 2020

The Municipal Plan, our most important planning document, sets out the building plans and the strategy for the next four year period. The plan has two main themes: standard of life and the environment, and it lays down one important principle to be followed in the city's development: Bergen shall be developed within the present city limits. This means that development must take place while preserving the present greenbelt and agricultural areas, as well as gardens, parks and recreational areas.

A public transportation network of a high standard will be developed, and this will be the city's most important "nerve centre." If we succeed in giving priority to residential building around the existing commercial and service centres, this will do most to reduce the need for transportation. Since these important issues are contained in a rather dry document, what can we do to make them visible to the person in the street and make them a subject of discussion?

Having a detached house outside the city centre is a status symbol, but living in the city itself has also become attractive. How can we show that new commercial and service centres can be developed into viable local communities, "small urban societies," planned to allow people to walk or cycle to or from the surrounding residential areas, and with good public transit to the city centre and other parts of Bergen?

We gave five architects the job of presenting their vision of how the most important suburban junction points could be developed into attractive built-up areas with a good local community. We also asked them to show the development in phases, for the years 2000, 2010 and 2020, respectively. Planners and selected individual experts at the city planning office supervised the work and made suggestions as the work progressed in order to make the architects' vision of the future as realistic as possible.

One of the sites for major redesign, Åsane, is an urban neighborhood which in the course of 25 years developed from an agricultural area into scattered housing for more than 30,000 people. Today, the area has direct linkage to the city via a motorway which splits the area into two sections with 15,000 people on either side. There are two pedestrian bridges 800 meters apart, the only connection between the two areas. The existing shopping centres are privately owned, and so are most of the parking places. Our idea was to cover part of the highway and develop a small town in this area, partially on top of the motorway, with houses close to commercial and service centres, work places and cultural facilities, and with areas for recreational purposes. Inhabitants of all ages would then have a local centre – a place with its own identity and offering a broad range of facilities.

We planners asked ourselves what we could do to re-create the original atmosphere and old landmarks. There were two fundamental elements: First, there was the old postal road which had gone through the area for hundreds of years, with old cultural landmarks still visible. Then there was a river which used to cross the area, but which had been re-routed through pipes long ago.

In the Architect's vision for 2000, we imagine building houses on the large parking site and establishing a new main road through the area. Parking places will be located under the buildings and a new multi-story car

park will be built at the entrance to the area. In front of the church, where the new transverse walkway and cycle path is to go, there will be a park with a pool. The river will be dug out and its channel will be a main feature of the new main street.

By 2010 the motorway will be lowered and covered by a 230 meter "lid." There will be outdoor sports facilities, as well as a new clubhouse and cultural centre on the other side. At this point, the area will really start to develop as a cohesive unit. A local railway line will connect the suburban area to the city centre. And by 2020, Åsane will have its own town centre, and the area will be further developed.

Now we shall move on to consider the centre of Bergen where a gigantic motorway has ruined a large part of the city centre. The historical harbor of Bergen, towards the north, has always been the first thing to greet visitors to Bergen, but it is no longer the gateway to the city. Now, people come by car and plane from the south, arriving through the "back door," and their first meeting with Bergen from this side is not particularly inviting. The land area of the highway interchange, parking and railroad tracks is equal to the land area of the entire old downtown. The five architects produced drawings showing how the big "spaghetti intersection" can be partly dismantled, and the traffic planners have con-

Freeway News

At present, freeways and railroad yards dominate Bergen's "back door," covering an area almost equal to the whole downtown.

Bergen, if the plans for 2020 are actualized.  The "new front door," the highway and
train entrance to the city, has been very substantially modified.

Jerri Holan

Bergen's old town today. New projects must support the historic pattern of compact development.

firmed that the proposed solution is acceptable, even with the same volume of traffic as today. This new approach opens up the way for new development potential in the centre of Bergen, freeing space for a beautiful new city area, with attractive residential qualities.

The new proposal will also include a new harbor on the south side of the city. Here, too, water will be the element which connects the various parts of the plan together – which is also in keeping with the historic tradition of Bergen.

Part of the work done has involved making a climatic analysis, taking into consideration air currents down to the neighborhood level, to see if the areas around the waterfront are suitable for residential accommodation. It turns out the present conditions are are not good in terms of habitability, and so, the proposal involves establishing a major greenbelt area to make it possible to combine healthier air with recreational facilities, providing a direct link with gardens and natural park areas further up towards the surrounding mountains.

We are already in the process of taking the first steps, which we believe are the most important ones to make the waterside area attractive, with the construction of walkways and cycle paths linking the waterfront area to the city centre.

We realized, however, that public discussion is crucial and our proposals for changing Bergen became an important issue in the recent election. To clarify, we mounted a major environmental exhibition and made the material available in printed form on application to the Bergen City Hall.

When the votes were counted, the winning political coalition had set clear environmental objectives for the next period of the City Council, and for reduction of $CO_2$ emissions by 20% by 2005. Immediately, those of us in the planning department wondered if the politicians realized what they were promising. Will we be able to deal with all the individual matters put before us in such a way that these promises are reflected in everyday life? This will be a great challenge for us!

What further action can we take to back up our vision of Bergen? In one area, the commercial sector and organizations representing special interests wish to come together and discuss a real strategy for the development of their local centre. We will proceed in much the same way, with workshops hatching out ideas and presenting the best in ongoing discussion. People from all the various interest groups will be represented. The task of the local authority will be to provide a framework within which the many issues can be discussed and clarified. Most important of all, the authorities must ensure that the individual phases of the development are guided towards the creation of a more appealing, healthy and altogether better local environment.

We have continued on in our ecological planning for three other parts of the city. In Nordnes, we worked with schools and parent associations to establish a new and greener school area with a school garden accessible to the

# Check List – Ten Environmental Goals for Bergen Planning

1. **Transport – location of functions**
   Various functions should be located for a minimum of transport need.

2. **Site and ground conditions**
   There shall be legal limits of health damaging radon, magnetic fields, noise and air pollution. Developments shall be planned in relation to landscape/topographical conditions and climate.

3. **Good social environment and high quality of life**
   New and old dwellings shall have a balanced social composition and shall be economically and physically accessible for as many people as possible.

4. **Public participation, local organization**
   Present and future dwellers shall have a genuine influence on the development and management of their own dwellings and neighborhood environments.

5. **Greenbelt, ecological totality**
   Housing estates, called public-assisted housing in many countries, shall have a continuous greenbelt and be in ecological balance. Roads, pedestrian walkways, parks and gardens shall be organised for children, young people, the elderly and disabled.

6. **Water, sewage, refuse**
   The consumption of drinkable water shall be reduced. Sewage shall be discharged in a manner harmless to the environment. Refuse shall be reduced and separated at the source.

7. **Energy consumption**
   Housing estates shall consume as little energy as possible. The dwellings shall make use of renewable energy resources.

8. **"Healthy" materials, indoor climate**
   Buildings shall be constructed from materials that are healthy to produce, use and to dispose of.

9. **Safety, security**
   The housing estates shall be safe both within the dwellings and outside, but at the same time not be overprotective towards danger.

10. **Aesthetics**
    Design shall encourage tradition in building, promote a healthy mind and give meaning, continuity and historical connecion.

---

physically handicapped. At Melkeplassen we remodeled 140 units of housing with better heat and sound insulation, more convenient access for the elderly and new traffic safety measures. At Regnbueåsen we developed what we called an "ecological housing area" working with a number of small developers. A public nursery went in immediately and we tried out new ideas and technologies for roads, surface water management, refuse management, drainage and plant-based water purification. To round out five years' work on these projects, we celebrated publicly with a major exhibition and special series of seminars in September, 1995.

Our next steps include continued follow up for "A Vision for Bergen in 2020," a study on $CO_2$ emissions for Bergen and an effort to ensure that new environmental objectives contained in "Healthier City 95" are incorporated in new legislation and by-laws and that new working tools are employed in administrative decision-making. In actual fact, planning does not have a high priority in Norwegian society, and many people regard town planning as unnecessary, and believe that it only slows development. It is therefore my view that one of the most important tasks which must be addressed by those of us who work in the public sector is to create a greater awareness that we are working actively for the common good – and that our efforts to plan and achieve a sustainable social development can only succeed if the municipal authorities are fully involved in this process.

Tohono O'odham woman and children in front of their ki, Sonora Desert, North America.

*But what is the wholeness of the city? Unlike the human body, where wholeness has been defined by natural evolution and our task of learning is mainly to respect that wholeness, the city is completely created by human beings. If we are to have good cities, as whole and organic as our bodies, is it not necessary for us to somehow devise a kind of shorthand for the millions of years of evolution? That shorthand is theory. Theory goes to the roots of form and function, means and ends, cause and effect, value and action. Theory, including design and experimentation, is society's substitution for millennia of accidental mutations. I doubt we can devise an organic urban system, or any system at all, without first creating sound and penetrating urban theory.*

Kenneth Schneider, ON THE NATURE OF CITIES

# 3. Ecocity Theory – Conceiving the Foundations

We need a new study, art, science and discipline for ecological community design and building. We knew some of it once, have forgotten most of it by now, but can discover, learn and invent afresh. Ideas are the most powerful tools we have.

In this chapter, Paul Faulstich digs into the evolutionary foundations for our physical communities and finds "geophilia" lurking there, now concealed from our consciousness by phobias upon addictions, technologies upon literal walls, sealing us off from our essence that yearns to breathe free. "The vast majority of the sensory experience and information that we process is of our own manufacture," he reminds us. (I remind us that we teach how to build, but what we build teaches us how to live.) Ecological design of our communities, he maintains, can liberate our authentic selves and revitalize nature. Marina Alberti digs into two of the four primary dimensions of human disaster on Earth: overpopulation (P for Population) and overconsumption (A for Affluence) – the other two being impacts of specific technologies (T for Technology) and the disfigured anatomy of the built community (L for Land-use/infrasructure). She uses the not-quite-complete formula of her colleagues at Stanford University, Paul and Ann Ehrlich $I=P(L)AT$, explained earlier in these pages, but which, if complete, is a clear formulation of humanity's presently dysfunctional role in the biosphere. Rusong Wang reveals the ancient roots of ecological community design in China, which will prove to be news to most people in the ecocity and ecological design movement, due to the unfortunate lack of communications during that long cold Cold War. Anders Nyquist discusses the ecovillages he has been building in Sweden, while explaining the conceptual foundations of his work in the ecocycles of nature. I portray ecocity mapping and zoning as a tool for village, town and urban transformation and as the first step toward an economy in balance with nature. And Marian Zeitlin traces trends that are presently in place that are leading directly – do not stop at government planning – to the ecocity and ecovillage future. At least the trends are there to be worked with, whether or not we take advantage of their positive possibilities.

This is a lot! How to build the ecocity? Here it is, in its leanest outline. Theory is not just theory – in the sense of any old idea about the rules of organization of whatever it is you are talking about. Here we are trying to get at "good," "correct" or "true" theory, that once found, becomes an organizing principle for understanding and action in our universe, a tool to improve our lot, a tool to participate fully in the normal, healthy and often surprising unfolding of the unknown future into the imperfectly understood present. With good theory we can repair our human-built home that is in such disarray, and deflect it away from assaulting and toward assisting life systems on Earth.

# Notes on a Natural History of Social Living – An Evolutionary Design Perspective on Reconnecting Culture and Nature

## PAUL FAULSTICH

*If we align ourselves with the spirit of place,*
*we will find humility infused with joy.*

– Terry Tempest Williams

Culture and biology are not mutually exclusive. Humans have an innate need for contact with nature; we are genetically designed to affiliate with the natural world. Nature helps us fulfill emotional and cognitive needs, and is an essential component of our humanity. Ecological design entails working within the logic and patterns of nature, and is one way to reintegrate ourselves with natural systems, thereby enhancing our humanity and reducing our ecological impact. By designing with Earth in mind, we reconnect culture with nature, and we recollect our essential affiliation with the natural world.

Evolutionary and ecological perspectives can, and indeed should, inform and affect the concerns of architects, planners, and developers. Many of our current patterns of design are socially and ecologically dysfunctional. They alienate by denying our cultural and biological needs for affiliation with nature. A deeper merging of the fields of human ecology and environmental design provides some hope, and helps us to reestablish, and to increase, our compatibility with the Earth and with each other. By combining a keen understanding of natural history with cultural aesthetics, ecological design enables universal needs to be uniquely expressed.

While my immediate focus is on broad aspects of design, I will present an implicit critique of industrial planning, food production, architecture, and transportation. Combining ecological design principles, social concerns, and evolutionary theory, I will explore alternatives to current (industrial) patterns of social living.

Designing for, and with nature, should be part of our contemporary quest for meaning, and of our quest for understanding our place in the world. Ecological design brings an elemental awareness of natural processes into

*Paul Faulstich is on the faculty in Environmental Science at Pitzer College in Claremont, California.*

our cultural lives. It embraces the premise that nonhuman nature is as important as humanity itself. Indeed, nonhuman nature is essential to our humanity, for it is our relationships to nature's various components that make us human. Ecological design accepts that for our ultimate fulfillment (and survival), humans must be able to accept natural constraints on our freedom of will.

**Evolutionary Design**

As animals, *Homo sapiens* have evolved in dialogue with place; our patterns of social living have developed through discourses with the land. Throughout our evolutionary journey (save for the last blip of industrialism), humans, and our dwellings, have co-evolved with place. Society, mind, and culture are all part of biological evolution; by logical biological extension, our physical constructions should reflect our evolutionary heritage.

Bioregionalism has become a contemporary framework for enhancing and understanding human relationships with landscapes. But it has emerged only in response to the chasms we have carved out that separate ourselves and nonhuman nature. Bioregionalism is the purposeful and conscious movement to inhabit specific places in a meaningful way; to learn the geology, climate, flora and fauna of particular biotic communities and to live with sensitivity to place. It is a conscious, ethical, and active expression driven by our innate need to emotionally affiliate with land; an impulse that I refer to as geophilia. Geophilia is somewhat different, and more fundamental than bioregionalism; it expresses tens of thousands of years of evolutionary encounters with landscape. While it is related to bioregionalism, geophilia just might be inscribed in our strands of DNA. As part of the ecological history of our species, geophilia exists today as a sort of collective memory of experience associated with the natural environment.

We carry genetic material that is Paleolithic, and through which we find the landscape and its component features compelling and essential. We need nature, not only as resource, but as cognitive sustenance. The way we design our communities and dwellings, then, should attend to this need by reflecting our ecological selves, and resonating with our full identities as cultural animals. Not only should design be responsive to resource limitations, but it should be responsive to our cultural needs.

Part of our humanness derives from the unique ways we affiliate with the land. Landscape is a critical element of human fulfillment at the individual, cultural, and species levels. Whereas the natural environment sustains us physically, the landscape sustains us bioculturally. Ecological design provides environmental insight, and allows for the comprehension of landscape and its ecological processes. As an organic expression of our intercourse with nature, ecological design facilitates human communication and societal distinctiveness. It enhances commitment, promotes ecologically based subsistence, and encourages ethical behavior and action.

Our affinity for wild nature is innate and integral to our ontogenetic and phylogenetic development. Designing with nature requires attention to, and appreciation for, the transhuman world. We begin the ecological design process by asking how we can create meaningful relations through the structures that mediate between ourselves and the rest of nature. We ask whether our present cultural structures (physical, social, or symbolic) fulfill basic human and ecological needs; we ask which are beneficial for transhuman life, and we consciously and sensually move toward an appropriate paradigmatic shift.

Place-centered dwelling embodies a human dialogue with the land and its nonhuman inhabitants. Through it, we envision ourselves as co-inhabitants of place. A crucial dimension of biocultural design is human, ethical, and geographical. Part of what it means to be human derives from careful reflection on the natural history of place. It is, then, essential for us to regain a notion of ourselves as extensions of the land before we can hope for a recovery of ecological living.

Evolutionary and ecological systems function at many scales and levels of wholeness simultaneously, from the metabolic dance of cells to the vast cycles maintaining watersheds or atmospheric patterns. By designing with nature and working within these natural patterns of emergence, we aspire to create designs that are compatible with the world, and hence compatible with ourselves. Evolutionary, ecological design is design for diversity. It is a socionatural process, recognizing that human cultural diversity is correlated with biological diversity.

## The Post-Industrial Pleistocene Possibility

Natural complexity is counterpart to human intellect and creativity. Most of our current patterns of living, however, find us increasing technological complication while decreasing ecological complexity. As we do this we are exploiting a phantom carrying capacity; temporarily ignoring our net drain on the Earth, and basking momentarily in the exuberance of an artificial plenitude.

To understand our contemporary design aesthetics (many of them ecologically dysfunctional), we have to recognize their roots – roots that inevitably have earth clinging to them. As others before me have proposed, let us re-think our buildings and villages as organisms, or perhaps membranes, and landscapes as extensions of ourselves. Let us accept the challenge and begin to re-design our dwellings and our communities for sustainability. But there is more to it than this. We should strive also to create inclusive designs, wherein humans are reintegrated into natural systems, back into the society of nature.

As sensual and cognitive creatures, we need complex, complete ecosystems; not only as terrain, territory, and resource, but as cognitive sustenance. We need connection with wildness; we are thinking animals that need to think of animals and the lands that sustain them. There is evolutionary advantage to emotional and intellectual affiliation with nature; just as we need love of humans to enhance commitment to families and friends, we need love of nature to enhance appropriate subsistence.

Being human doesn't exempt us from ecological principles. One of the problems with evolutionary theory as it has developed is that it emphasizes competition at the expense of cooperation. In the natural world, evolution and selection are not so much about survival of the fittest, as about survival of the fitness; how things fit together, as Dolores LaChapelle describes in LISTENING TO THE LAND: CONVERSATIONS ABOUT NATURE, CULTURE, AND EROS. Ecosystems are just that; ecological *systems*. They function as networks of relationships. Our dwellings and communities should – indeed must – reflect similar cooperative systems.

In post-industrial ecological design, we acknowledge that the human is derivative and that place is primary. Most people from industrial societies live in an ecologically insular and provincial way. We – and I speak for myself here – are isolated from many of the very forces that shaped our evolutionary being. This is profoundly problematic; the vast majority of the sensory experience and information that we process is of our own manufacture. Most of the sensory information we receive is fabricated, and mediated by machines. Much of what we encounter, consequently, is a human artifact; an artifice of the natural. The social, evolutionary, and ecological consequences of this are proving to be immensely damaging. But there is hope, and ecological design is going to be part of the solution that will carry us, meaningfully, into the next millennium.

## Socionatural Design

Humans are not suprabiological beings. Our cultural identities, and our very existence as a species, are dependent upon our organic dance with nature. We are products of our associations; human brains need encounters with the Other. We need to interact, not only with other humans, but also with the winged, the four-legged, the burrowing – the rich array of life that informs our humanity. These interactions supply more than metaphors, they supply stimulation, companionship, and a nexus to our wildness. They demand we be attentive, and they provide models for the organization of human social living. We are, quite literally, *of* the world, and our species' ecology is the nature of our organic relationships.

Encoded in our biology is an affinity for nature. The core of ecological design lies in the recognition that this affinity is fundamental to healthy patterns of living. In developing an ecological design aesthetic, we should base our considerations on the biological needs shared by all living nature. An ecological design aesthetic finds compelling anything that enhances coherence, renders the landscape comprehensible, contributes to a sense of community, and honors natural patterns and processes. To recognize the essential characteristics of nature is to reveal a working vocabulary for our design processes. Nature is the grammar of our aesthetic language; it makes our own contrivances comprehensible.

Ecological design, then, involves critical reflection on natural forms and rhythms; the Earth and its processes must inform and give form to our works. We are not, for example, adapted to live at temperatures or lighting levels that are uniform and constant; we are most perceptive, most responsive, most *alive* when we experience variation in our surroundings. Sensitive design incorporates change at many levels; diurnal, seasonal, generational, emotional. It produces structures and designs that are not static, but are responsive to our needs as social animals. These needs are *of* nature; they are transhuman, interdependent, and innate. Designing a dwelling that is efficient and healthy also means designing a dwelling that is responsive to natural processes; ecological design merges natural history with social living.

Ecological design is not simply about constructing. It is also about carefully and liberally choosing what *not* to develop. Wild lands are certainly not of human design, but in ecological planning they exist as design considerations. Wild nature is the core of our beings; it is the megacommunity from which we emerge. The preservation of such wild places is a central consideration of ecological, evolutionary design.

From an evolutionary perspective, design aesthetics are not to be measured by some culturally constructed notion of beauty, but by ecological considerations. Beauty then comes to embody the regenerative, organic,

Zuni Pueblo in New Mexico, illustrating the association of housing and public open space and use of local building materials.

and wild attributes of a landscape. An ecological aesthetic provides a vision across boundaries, appreciates diversity, and senses basal connections between beauty and wildness. What is beautiful in a design is what is humble, healthy, and sustainable. Beauty is life-affirming; it resonates with the biotic community.

Scale, too, is critical. Hunting and gathering cultures, and other small-scale societies, participate in social and ecological systems that engage individuals in networks of relations. Their use of space facilitates the intimacy of small group, egalitarian decision making. The use of space in industrial complexes, by contrast, tends toward the impersonal and authoritarian. The scale is immense. In rethinking our use of social space, we can seek examples from societies wherein we see scale working to conserve biodiversity, to protect people from authoritarian rule, and encourage the engaged participation of a diversity of individuals. This is not to suggest that we

trivialize Native peoples by trying to reinvent the 'primitive.' We can't go back, largely because we never left; ours is the same biological heritage as our indigenous cohorts.

Ecological design should not be just about sustainability. It should be, also, about the full range of psychological and biological needs of *Homo sapiens* and our fellow cohabitants of Earth. These needs can only be truly met through a dialogue with transhuman nature. Ecological designers are not exactly creators. We are cultivators, or shape shifters. Like kneading multi-grain bread dough, we massage natural forms and materials together with cultural processes. The product is a hearty and healthy socionatural landscape.

# The Impact of Population and Affluence

## MARINA ALBERTI

"How many people can sustainably live on Earth, that is, without undermining the capacity of natural ecosystems to sustain the human population in the future?" This is one of the most complex questions of our time and one of major concern. Population growth is increasingly recognized to be the key cause of environmental problems. However the relationship between population, economic development and environmental change is too complex to support such a simple conclusion.

We can understand human impact on the biosphere as dependent on population growth, consumption patterns and the impact of technology (I = PAT). It is also clear that social relationships affect the interactions between population and natural resources.

Population trends and the scale of human consumption now dominate the biosphere and portend profound problems. At the beginning of this century world population was less than 2 billion. Today world population is 5.8 billion and is predicted to reach 8 billion to 9 billion in the year 2025. Ninety million people are currently added each year to the population of the globe.

Most people are unaware that already humanity has taken over Earth's natural resources and affected natural processes at unprecedented rates. Humanity uses directly and indirectly nearly 40% of the terrestrial Net Primary Production, accoding to Peter Vitousek, Paul and Ann Ehrlich and others, which is all the solar energy annually captured worldwide by photosynthesis on which life on earth depends. Most people are also unaware of the degree to which humanity has taken over Earth's land surface. 2% of Earth's surface is occupied by cities and towns. But much more surface is used to support the world population in terms of resources use and waste

*Marina Alberti, educated in Venice, Italy, is a member of the faculty of the Department of Urban Design and Planning at the University of Washington in Seattle, Washington. She was a post-doctoral researcher at Stanford University at the time of the conference. Here we combine points from both her lecture and her paper on population and consumption.*

sinks. 11% of the world's land is used to grow crops; 25% serves as pasture for livestock; and most of the 30% that is still forested is exploited at some level or converted to tree farms. Nearly all the remaining 30% is in Arctic or in Antarctic regions or desert or is too mountainous to be used for cultivation for human purposes. In addition humans appropriate 30 percent of total accessible renewable freshwater. About 17 percent of vegetated soil has been degraded in the last 45 years. Nearly one-third of the planet's arable land has been lost to erosion in the same period. These figures have been researched by Sandra Postel in 1996, Leonard Oldeman in 1990 and the Food and Agriculture Organization of the United Nations (FOA) in 1995 and are considered accurate by many other scientists.

As population and consumption trends increase their pressure on the biosphere and resource base for society, there is an increasing gap between rich of the "North" and poor countries of the "South." With less than 20% of the total population, developed nations are consuming more than 75% of all natural resources including food and energy and are putting 75% of the pollution load on the Earth's capacity to absorb emissions and waste. Within countries too, disparity between rich and poor is growing. The condition of relative economic equity in Norway referred to earlier by Bente Florelius, is becoming an increasingly rare exception.

The population of developing nations is growing at a larger rate – 2.5% per year – compared with 0.6% per year in developed countries. The poorest are the fastest growing segment. And while certainly the condition of populations in many developing countries has improved, this is not true for all populations. Most people still live in countries where average per capita wealth is about a fifth or less than that of rich nations and where babies are some five to twenty times as likely to die by the age of one due to the conditions of life and lack of sanitation. Of those, nearly a billion live in absolute poverty – too poor, according to Ann and Paul Ehrlich to buy enough food to maintain health or perform jobs.

Susan Felter

Marina Alberti

There is a nutritional gap of almost 1000 calories per capita between developing and developed countries, the FAO reports in 1995. This gap is much greater if the calories lost in converting primary production to live-stock are taken in account. In sub-Saharan Africa the number of undernourished has increased both in absolute terms and as share of total population. In this region food availability per capita has declined below 1970 levels.

Environmental impacts range wildly depending on who we are talking about, that is, depending upon consumption levels and lifestyles. All other things being equal, more people generate more pollution. But as I have just described, not all other things are equal. One person added to the population of the North is not equal to one person added to the population of the South.

Charles Hall from the State University of New York has estimated that an American baby born in 1990 will, in its lifetime, produce about 1 million kg (2.2 million pounds) of atmospheric pollution, 10 million kg of liquid waste and 1 million kg of solid waste. In addition, he says, this baby will consume 700,000 kg (1,540,000 pounds) of minerals and 4000 barrels of oil. To compare the effect of an additional person in the North and in the South of the world we can use this average measure proposed by Hall. Of the additional billion people that will be added to the world population in the next decade, more than half will be in Asia, approximately one quarter in Africa and 10% in Latin America. Only 5% will be added in the USA, Canada and Europe taken together. However the total impact of this 5% will cause more environmental impact than the other 95% since the populations of these regions are consuming resources at rates 10 to 100 times higher than that of most of the other countries.

This is not to say that population growth is not as important as consumption patterns. This is to demystify the myth that population pressure comes only from the South. Certainly population is a critical multiplier. To give an example, as Paul and Ann Ehrlich have said, suppose humanity succeeded in reducing consumption by 5% and reduced the impact of technology by another 5%. Taken together, these would reduce environmental pressure by 10%. But at the current rate of population growth, if all else remains equal, it would take only six years to bring the pressure on the biosphere back up to pre-reduction levels. Therefore it is necessary to act on both sides, population and lifestyles.

Let me conclude with a few words on what changes can be made to help address this problem. The Conference on Population and Development in Cairo in 1994 was an important step in this direction and provided signs of hope. There are important aspects of the discussion that need to be addressed. There is no doubt that population growth (fertility rates) in developed countries has dramatically declined with human development, but this has often been over-simplified. Often fertility declines have been associated with increased national income. But recently several studies have been able to distinguish between the effect of income growth and other aspects of human development such as economic security, sanitation and nutrition, education especially for women and particularly women's economic independence and parity in society. Based on these recent studies and other information, the prevailing notion that population problems can be solved by providing industrialization and birth control devices to developing countries is demonstrably outdated and inadequate. This prevailing notion is thus clearly an imposition of western culture and values on the countries of the South.

The Women's Conference in Beijing stated forcefully that it is time to re-frame the approach to this problem. Instead of asking how to reduce fertility rates we need to ask how to improve women's lives and how to give them equal power in shaping the world future.

# Ancient and Recent Ecological City Theory and Practice in China

## RUSONG WANG

Dramatic changes have taken place in the world in the last two decades of this century. Conspicuous among these is the development of China, the most densely populated country in the world, and its rapidly growing impacts on the planet. During the past 12 years, the national production value of industry has increased by three times, and rapid urbanization is predicted to continue for many years. Over 500 million people will be living in cities in China by the year 2000 and their needs, together with the urgent demands for industrialization from another 800 million rural people, will exert extraordinary pressures on the environment and society.

---

*Compiled and edited from papers by Rusong Wang, who, on his way to Yoff, was unable to obtain a visa and use his air ticket to pass through New York City, due to the United States government budget negotiations impasse in December, 1995. He never arrived, but the following represents some of his recent work. Rusong Wang is Director of the Department of Systems Ecology, Chinese Academy of Sciences, Beijing, China and President of Urban Ecology China.*

*A little more introduction is in order. As Joan Bokaer has pointed out, in the debate on China's near future hangs the balance of the development patterns of much of the world. With China's 22% of the Earth's population swept away in stunningly rapid economic growth, the impact of China on world resources and the biosphere would be profound even if it were not to influence other countries' paths into the future, though obviously it will. Right now Chinese policy makers are trying to decide between rail/compact city development and highway/car city development. Recently-successful young individualist business people tend to favor cars and highways. More traditional professionals in planning, and certainly the kind of thinkers represented by Rusong Wang, favor a new way bred of both ancient tradition and new ecologically tuned technologies, policies and strategies for living. For most of us, Dr. Wang's perspective will be a surprising revelation of a strong and historically deep ecocity tradition that most of us simply never have heard about.*

Yet beneath the present pattern of rapid growth and change in China lies an ancient foundation for development based on ecological principles. In addition, there are a number of contemporary theoretical advances and new approaches, developed in actual city planning and administration practice, with very positive potential and a growing record of early successes.

For thousands of years Chinese philosophers have investigated the harmonious relationship among Tian (heaven or universe), Di (earth or resource) and Ren (people or society). The result is a systematic set of principles for managing the relationships between people and their environment, including Dao-Li (natural relationship with the universe, geography, climate, etc.), Shi-Li (planning and management of human activities, such as agriculture, warfare, politics and family), and Qing-Li (ecological ethics, psychological feelings, motives and values towards the environment). The Yin and Yang theory (negative and positive forces played upon and within all ecological relationships) also relates to understanding and managing human interaction with nature and so does Wuxing theory (five fundamental elements and movements within any ecosystem promoted and restrained by each other) and Feng-Shui (wind-water theory expressing the geographical and ecological relationships between human settlements and their natural environment). All these are parts of the ecological principles for planning society's works and our economic and cultural lives.

In the 21st century BC, when flooding was the most urgent human ecological problem facing the ancient Chinese and when various measures taken to protect flood waters from spilling out of the rivers were failing, the King Yu, reputed founder of the Xia Dynasty, successfully contained the flooding rivers by replacing the strategy of blocking the river's path by a dredging-through strategy to let the flood flow smoothly down to the sea in its own way, at the same time making the water serve the people. This is a good example of using eco-cybernetics to organize an ecological order and to harness the natural forces to serve society.

In 1984 professor Shijun Ma summarized theories and views on general ecological design (both natural and human-created) in the following four principles of eco-cybernetics:

Axiom 1:  Waste-product circulation principle: Every product in the world will inevitably become a waste; yet every "waste" is bound to be a "resource" useful elsewhere in the biosphere.  Too many or too few "wastes" will cause various ecological problems.

Axiom 2:  Positive-negative feedback principle: For every creature there will exist certain positive feedbacks and other factors to promote its development as well as certain negative feedbacks and limiting factors to restrict its development.  In a stable ecosystem, the positive and negative forces balance.

Axiom 3.  Competition-symbiosis coordinating principle:  All natural creatures survive through competition for resources as well as symbiosis for sustainability.  Those species lacking either competition or symbiosis mechanisms are weak in vitality and will eventually be replaced by others.

Axiom 4.  Exploitation-adaptation evolution principle:  All creatures use survival strategies of seizing chances for favorable development and avoiding risks in order to maximize self-protection.  A stable ecosystem consists of those species having strong abilities for exploitation and adaptation.

Systematizing all this in 1984, Ma proposed what he called "Social-Economic-Natural Complex Ecosystem," or SENCE theory.  He advocated carrying out eco-construction projects in rural and urban areas.  Major issues, such as food, energy, population, resources and environmental pollution, he pointed out, are all related to social systems, economic development and natural environment upon which human beings depend.  *(Wang became involved in this work originally as a mathematician trying to devise mathematical formulas for planning and predicting both social behavior and ecological patterns of change, only to*

Dixie LaGrande

Bicycles, like these in Beijing, are extremely low-impact vehicles.  Cars are now beginning to shoulder them off the Chinese  streets and roads.

*decide that both society and ecology were far too complex with far too many significant variables to be analyzed and manipulated dependably with math.  Put the two together – society and ecology – and the complexity becomes near infinite, and predictability and management for healthy results impossible through mathematics.  Obviously, he concluded, we need a new science or art of ecocity planning and design, a discipline that he decided to plunge into.)*

Although society, economy and nature are different in quality, said Sijun Ma, and each has its own structure, function and developing laws, each of them can benefit from and be restrained by the others.  It is clear that particular facors within this kind of complex set of factors cannot be considered separately.  From the complex ecosystem perspective, the cross-linked interrelationships between the subsystems, the dynamics of material, energy and information flows and the trade-off among benefit, risk and opportunity of the Social-Economic-Natural Complex Ecosystem (SENCE) are the keys to solving the major problems of our age.  SENCE theory undoubtedly points out a way for ecopolis construction.

"Eco-county" is the abbreviation of the planning, management and construction of a county-sized complex ecosystem according to ecological principles. Generally speaking, it is to build a vital environment of production and living within a county area, which has a high efficiency of resource utilization, harmonious relationship among its components, and self-regulating mechanism towards sustainable development through regulation of human behavior according to eco-cybernetic principles, i.e. regeneration and circulation, coordination and symbiosis, and self-organization and autotrophy.

An eco-county project was launched in 1986 in Dafeng County, China. New integrative production complexes were built up to enhance the symbiotic relationship among the different sectors, trades and regions. Niuqiao village was taken as a case study site to put the ecological planning into practice. The measures involved three aspects: l.) to rearrange the crop structure, such as planting forage, mulberry and fruit trees according to the farmland features instead of mono-culture structure of cereals and cotton, 2.) to change the family-unit production system into a village-unit system and set up five production serving teams according to the agricultural production process, and, 3.) to hold training courses to popularize the new technology and to raise ecological awareness of the farmers. After three years experiment, the village has improved the structure and function of its production system. Compared with that of 1987, the gross production value increased by 48.3%, the income per capita by 21%, and the organic matter content in the soil by 6% while the input of energy and investment did not increase.

As a result of this experiment, ancient tradition, new thinking such as Professor Ma's and a climate of more openness and experimentation in China, the People's Government of Maanshan City and the Research Center for Eco-Environmental Science of the Chinese Academy of Sciences entered into an "Agreement to Initiate a Project on Eco-City Planning and Construction of Maanshan City" in 1991. Small experiment projects were begun there in 1991 and 1992 on the eco-neighborhood design level, together with an assessment of required legislative change and leading to completion of an Eco-City Master Plan for Maanshan City by 1996.

Human beings have been dreaming of an ideal habitat or eco-city for thousands of years. Ecopolis study in China does not seek to build an ideal model for demonstration but rather to look for a feasible way and healthy process to achieve sustainable and harmonious relationships within its own system and with its environment. The ecopolis or eco-city envisioned will be permeated with vitality and health even though its level of material wealth may not be as high as that in richer cities that lack these ecological qualities. When the local people fully understand the dynamics and eco-cybernetics of their own cities and act spontaneously with nature, ecological development will be realized.

What is the ecological order on earth? It is a kind of accumulated energy described by H. T. Odum in 1983 as a kind of steering force or, in Chinese saying, Qi, a kind of higher quality of information or low entropy, and a kind of holy spirit. Neither an individual planner in a limited period and space nor a single discipline can see and understand it, just as a cell of the human body will never understand the function and developing course of the whole body. But a set of cells can formulate an organ, and a set of organs can formulate a body. The dynamics and cybernetics can be understood in these higher levels. So we need a new kind of integration, by which we should jump from our own position in the multi-dimensional niche space to a higher holistic position. We need to think and act world-wide and generations long. This needs a breakthrough of methodology. Three centuries ago, the breakthrough in physics had brought about the revolution of mathematics, which in turn stimulated the progress of physics. The breakthrough in human ecology in the near future will be bound to bring about a subversive revolution in systems science and feedback to itself.

Searching for and acting with the eco-unknown – this is the very task of human ecology.

# Ecocycles – the Basis of Sustainable Urban Development

## ANDERS NYQUIST

Anders Nyquist

Today's city is an unbalanced system which depends upon our use of cars and the automobile infrastructure we have developed over many years now. This version of the city is a linear system in terms of energy and nutrient flow; it draws continually upon its outlying rural areas for sustenance and returns concentrated pollution and waste.

But we can change this. We can learn from ecocycles in nature the connection between ecology and economy, thus discovering how to achieve sustainable rural-urban interactions. Our cities' energy use must be patterned after growth and metabolism in biological systems. Viewed in this way, present industrialized society is a tragic parenthesis in the history of human kind. Uniquely resource demanding, economic growth is based solely upon increasing gross national product. Interest rates on borrowed money to fuel continued production have become the new slavery throughout the world.

If we want to change we must do many things, and I suggest the following: 1.) limit the growth of the population, 2.) bring industrial production into line with nature's limitations, 3.) increase self-sufficiency with locally produced food, 4.) use renewable energy, 5.) clean our waste water in local ecocycle systems, 6.) limit the use of non-renewable raw material, 7.) invest in re-using goods and material.

In Sweden, we have begun to address some of these issues. The first paragraph of the Swedish nature resource law reads: "From an ecological, social and public economy point of view, the ground, the water and the physical environment in all, shall be used so that a long-term good economizing is promoted." From this, we have coined the term "eco-village" to describe what now number 35 community developments in Sweden where inhabitants actively work together with architects and planners to design and take responsibility for their own

*Based on a talk and a paper by Anders Nyquist, architect and founder/designer of ecovillage Rumpan in northern Sweden and architect for many other constructed buildings and community projects.*

domicile from an ecological view point. It is more than just a living place; common production of goods and services is included in the design.

The model I use for eco-village design is the old traditional farmers' village with its close human relations and the fundamental idea of keeping the balance between the people and nature. I then integrate established technical solutions within the ecological field, such as passive solar energy and local supply systems. Also important is the incorporation of humane governance and management systems.

Rumpan, the small village where we live, is called the first model eco-village in Sweden. My wife, Ingrid, and I started it thirty years ago when we bought an old farm of 18 hectares (44 acres) close to the Baltic Sea. We started by writing down how we wanted to live in this hamlet or village before we started the planning procedure. Then we stuck to these ideas without compromise whatsoever.

We created the site layout and worked with the local authorities to complete the necessary paperwork. We started on site in 1967. We began by restoring the old farm land, building roads, developing the water supply, and building the first three houses. Since 1968 we lived there on holidays and weekends. Today we have about 25 families in the village. It is a three generation village with all kinds of people living there. The village has a tenant-owner society which is responsible for the following:

1. roads within the village,
2. communal cultivated and forested lands,
3. water and sewage,
4. energy distribution,
5. waste disposal and recycling,
6. maintenance of common facilities and buildings,
7. cooperative day-care nursery,
8. building future workshops.

We have a rotation system for the Board, so every adult has been in a responsible position. Every summer we have a meeting where we make decisions for the next year. Throughout the year, we do a lot of work together. That's the best way to learn from each other, and we do things cheaply that way. It gives the children opportunities to experience different things during the year. We have big parties together, too.

Three years ago we moved out to the village to live there year-round in a new prototype ecocycle house we built where we even run our own business. We used ecocycle technique in order to get experience so we could advise our clients how to build in the ecocycle way. We have separating toilets and small scale waste water treatment, which we ultimately reuse; it is as good as fresh water. We wanted a low energy consuming house, so we made it earth covered to protect it from the cold winter climate in northern Sweden near the Arctic Circle. The sun provides heating energy and firewood, burned efficiently, supplies the remainder of our heating needs. The supply air is heated or cooled as it passes through pipes in the earth and the whole building is like a heat exchanger. Local materials and good paint with low emissions are chosen. Most of the elements are locally made and screwed together so they can be re-used. It's an inexpensive house with low yearly costs. The house is

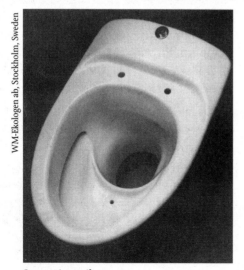

WM-Ekologen ab, Stockholm, Sweden

Separating toilet.

about 135 square meters, or 1,250 square feet. It costs $200 US per month to own, operate and maintain.

We built our greenhouse from recycled materials taken from the city's garbage, and we use it all year around. We are growing our plants in sand covered in fresh cut grass. I pick tomatoes in winter, when it may be -20 centigrade outside. We take care of our own waste through earthworm composting in our greenhouse.

We designed a country school in the eco-community of Timre in the north of Sweden. The school is built of natural materials with low or no emissions. Its energy needs are supplied by local renewable sources, including passive and active solar energy systems. Simple construction, screwed together, can be re-used. The electric installations are encapsulated. Rainwater is collected in a soil infiltration system. Separating toilets are used, with feces being composted in the building, and urine going to the farmers as crop nourishment. Food is cultivated in greenhouses around the school. The design utilizes natural ventilation and daylighted rooms throughout the whole school. The school's ecocycle systems, materials and constructions serve as elements for environmental education in the school. Experience gained from building this school will be incorporated into all the other schools in the community as they are renewed.

We are also applying ecocycle design to some larger projects. One is an Agenda 21 project near the city of Helsingborg. It is one block of land with about 500 apartments of older housing stock. We are going to try ecocycle methods to renew the old area. Another project is new housing close to the city center of Helsingborg. The blocks of land will incorporate the same ideas for ecoycling already presented, but here they will be employed in the middle of the town.

The importance of utilizing local, small-scale ecocycle systems is highlighted in another project we have started, called Engeshojden in Jander. Two years ago we were among six contractors invited to participate in a competition called "Living Close to Nature." The area to be developed is close to the Baltic Sea. It sits 10km from

Anders Nyquist's ecocycles diagram illustrating the integral connections of various aspects of the ecologically healthy homestead or village and its agricultural systems.

the city center of Yedna, north of Stockholm, and consists of 43 new houses in an old fishing village. The most interesting problem we had to solve there was water supply. The ground water is brackish, and fresh water is not easily available in the surroundings. We are going to re-use the waste water in this project. The waste water, rain and drainage water is sent down to ponds where plants and sand cleanse the water. It takes one and a half years for the water to move from the pond's inlet to where we draw out clean water and recirculate it. This ecocycling system makes it possible for snow and rain to provide adequate water for the project. The work on site will start this summer.

Now you may ask why separating toilets? Human beings each produce through urine 6 kilogram of nitrogen, 1 kilogram of phosphorous, and 1 kilogram of potassium annually. 500 square meters of cultivated land can support 75% of a person's nutritional needs, with an annual input of about 6 kilogram of nitrogen, 1 kilogram of phosphorous and one kilogram of potassium. So why not use it? We are self-sufficient in that way. (*At this point the audience claps enthusiastically*.) We all are sitting on a fortune! (*Now the audience laughs and claps uproariously*.)

Of the nutrients that leave the body, 80 – 90% is in the urine, and urine is sterile. We can neglect the 10 –

20% that is in the feces, because there are bacteria and viruses in the feces. The bacteria can live for about 3 months, and they can be killed through treatment. However, the viruses can live from 3 months to 24 months and cannot be killed reliably through conventional sewage treatment. Feces can be composted in an earthworm box – the earthworms eat the viruses. Then it can be encapsulated under a layer of soil and used to feed things like flowers. Lime can also be used to kill viruses; they die at pH 12. Feces also can be dried and burned; it has the same energy content as birch firewood.

We must change our old fashioned water system, which is based on 2000 years of misunderstanding, a system going back to the Roman cities with their aqueducts and sewage canals. Our waste treatment is based upon the erroneous belief that if you dilute the waste until you cannot see it, then it is OK. Every day 50,000 children die from diseases in water caused by bacteria and viruses. During this conference week, more than 300,000 children will die while we are discussing what we ought to do. Let's face the facts. Let's start to change the system and move to an ecocycle system for the future of our unborn children.

# Ecocity Mapping – Physical Structure and Economy in the Emerging Eco-Communities

## RICHARD REGISTER

Built communities, from the hamlet scale to the large city, have an anatomy. They have living bodies, of which our bodies are a part. The community anatomy can be represented in two dimensions as maps of the basic land-uses and infrastructure, that is, the major features of the community's buildings, streets, parks and natural features such as creeks and hills that emerge through the houses, offices, shops, yards and other structures. Usually, areas immediately adjacent town, whether nature, agriculture or bordering towns, are also featured on maps of our cities.

Zoning maps in particular describe what is *supposed* to be where according to size and shape of buildings and their uses – as allowed by the imperfect consensus process of the citizens – or at least the citizens with influence. The people whose consensus the zoning maps represent are various mixes of the wealthy business people, the powerful politicians or the well organized workers, neighborhood organizers, poor people needing housing, architects and artists wanting to make a living and contribute to the town's beauty and so on.

Now we need a consensus that makes it possible to build the ecologically healthy community, and ultimately, a creative, compassionate civilization. All of us who care about our communities, nature and the future should be in this group. Hopefully, this means all of us.

It happens that at this point in history our cities are either literally this or rapidly transforming toward this: sprawl development, automobiles, freeways and oil-based technological infrastructure. More specifically, we see large scattered areas of generally single use development with large generally single use buildings in the centers supported by and completely dependent upon a colossally large automobile, asphalt and gasoline infrastructure including parking lots, streets, gas stations, car washes, auto supply stores and garages, auto insurance agencies, traffic courts, gasoline cracking plants, oil pipelines and ocean-going tankers, automobile fabrication and assembly plants, oil and car businesses and Exxon executive offices and on and on.

Without much elaborating here I will say that it is a disaster to take your breath away, killing a quarter million people a year outright in accidents, exhausting in a couple hundred years the fossilized chemical energy it took the Earth a couple hundred *million* years to accumulate, covering much of the worlds best agricultural lands and beautiful natural landscapes, polluting water and air simultaneously both locally and globally, changing climate, and playing a leading role in the extinction of species world wide. The larger body we are part of *is* this tragically disfigured body of the built community. I call this disease autodysformia, a distortion of the form of the community by automobiles and all the attitudes and economics that make the automobile society so dominant on the planet today. Autodysformia is "auto" also in the sense that it happens automatically, that is, without much thought on our part. Our society and economy dysfunctions so badly largely because its physical body has grown into a spread out mess. As form matches function and anatomy matches metabolism strictly everywhere in living systems, both biological and ecological, the sprawled form of the city matches its destructive function precisely.

If the built human habitat were healthy it would be something else: the compact pedestrian community of very diverse uses, with almost no cars, supported by bicycles and transit and powered by renewable energy sources such as sun and wind. It would be, as Jeff Kenworthy says, the "walking city," of which some parts of existing cities are already good examples. Whether small scale or large, it would be the "ecocity" or "ecovillage" covering a small fraction of the land consumed by today's towns. It would be tall and small, that is, essentially three-dimensional rather than two, and small enough to be traversed mainly by foot. The principle here, which applies to all living organisms, and generally to their environments, is "access by proximity." Design communities so that, instead of having to drive "over there," what you need or want is just around the corner.

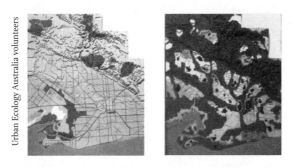

Urban Ecology Australia volunteers

"Shadow Zoning," as applied to Adelaide, Australia, produces concentrated island cities, towns and villages surrounded by restored rivers and natural and agricultural lands. These maps are projected from the present forward 150 years.

Since we must think in three-dimensional terms now and since "uses" usually portrayed in maps would be arranged one over the other as well as side by side in the ecologically healthy community, we need to combine the notions of both land uses and infrastructure. The infrastructure is the buildings themselves and their connectors, their systems of supply and waste, their streets, vehicles, bridges, stairs and elevators, their pipes and wires, their recycling and communication systems. The present relatively flat land-use/infrastructure includes freeways, overpasses, parking lots, garages, gas stations, the cars themselves..... What we are describing here, again, is the anatomy of the city. I suggest we make a word for this anatomy and call it "landustructure" since "land-use/infrastructure" is long and awkward. Knowing that we actually have an anatomy here makes it possible for us to much more clearly describe it and its functions, as the terms of the anatomy of the human body make it possible for the surgeon and the artist to go about their life-saving and life-enriching business.

Early in this conference Tukumbi Lumumba-Kasango suggested that we need to design a whole new economics to solve the vast imbalances of the present economy, both in terms of social equity imbalance – disparity between the rich and the poor – and in terms of humanity's imbalance with nature. But if a new economics is needed, what in fact would it build, what structures, technologies and services would it actually *be*? How would we go about designing it and building it?

To modify the existing structure, even in ways that appear positive at first glance, such as building more energy efficient cars, often turns out to be worse, not better. "Better cars" are most "efficient" in encouraging people to drive farther, and thus in a few years, sprawl and automobile dependence is far worse and more cars are required to cover the landscape and collectively more energy and land is being used and more destruction results. This is not theory or extrapolation from trends. It's history. From 1973 to 1993, years that start with the Oil Crisis of 1973 and end with far more efficient cars

than those at the beginning of those two decades, California's population increased approximately 50% while the area of its cities increased approximately 100%

If we pursue this line of thinking very far we realize that we need to have a strategy for transforming existing cities and this strategy needs to start with a clear conception of the "landustructure." On that foundation of healthy urban anatomy, a new kind of mapping to represent it through changes into the future can be devised. If we can't draw such maps we can't know what it is we are actually going to build or what its functions – or dysfunctions – are likely to be. Here's my suggestion from working in my own community for more than twenty years: start with an "ecocity zoning map."

Producing this map progresses something like this: First locate your town in time and place by understanding its basic history, climate, flora and fauna, as the bioregionalists advocate. Then look at your city, town or even rural village and locate the centers of greatest activity. Imagine adding more development there, but development that complements the existing uses with those that are missing or under-represented – all within a walking or bicycling distance. Remember: access by proximity. *Don't* provide new parking. In locations far from centers, where dependence upon the automobile is greatest, zone these areas for restoration of nature and agriculture.

If you do this and you allow for zones of decreasing density as we move out from the centers, you will produce a map that looks something like a pattern of target-like concentric circles around your town's centers, larger targets for the major centers, smaller ones for the neighborhood centers. In the zones farthest from the centers, we are actually calling for withdrawal from sprawl: "depaving" (tearing up concrete and asphalt paving) and moving or demolishing and recycling of run-down buildings to create space for nature and agriculture. Large, sprawling suburban areas can also find their centers and withdraw toward them, creating vital real towns surrounded by nature and agriculture, each with its own viable economics. Small ecovillages, probably in most cases supplying food, wood products, fish and other items to the larger centers, as advocated by Joan Bokaer, while receiving tools, other supplies and various cultural items in exchange, could condense like bright points of activity in the dissolving suburban fringe and appear as new villages in a landscape of industrial monocropping transforming toward a finer-grained pattern of farming and natural areas.

In such mapping we are beginning to visualize and record for others exactly what we are talking about. That's the first step toward a new economy, knowing what it is you want to build and making it available for other people to know.

It turns out there are only four basic steps toward this new economy. Ecocity mapping is the first. The second step is, now that we know what it is we need to build, how it functions and where it is located, drawing up a list of technologies, businesses and jobs necessary for its creation and maintenance. Instead of scattered single-use office buildings, shops and homes, cars and freeways, for example, we will need to build complex buildings, with renewable energy hardware, streetcars, rails, stations and bicycles. Instead of massive quantities of wire and pipes for communications and sewage services, a small amount of those things would be required – while much more compact hardware and service for information flow and recycling of human and other "waste" would be required. Lots of glass and gardening supplies would be needed for solar greenhouses and rooftop gardens. Elevators and ramps would need to be built. Car dependent buildings in the wrong places with little or no historic or esthetic value would need to be demolished and recycled, and some better buildings would need to be moved. Jobs and services for all of the above and much more would be required.

The third step in our "four steps to an ecology of economy" process: we need to write incentives so that people can make a living at and volunteer in support of the ecological rebuilding process. We need to make way

legally for the new zoning, raise height limits toward the centers (generally) while requiring more diversity – housing, for example, in and adjacent business, education, shopping and entertainment areas. We need to change land tax law so that over a few decades it becomes progressively more expensive to build in car dependent areas and more profitable to build in and adjacent city and neighborhood centers. Instead of subsidizing automobiles – quite covertly for the most part today – we need to eliminate those subsidies and gradually bring support to all aspects of the ecological rebuilding enterprise.

For the first ecological communities of modest scale, for example towns that might replace military bases, there needs to be an advertising and recruiting drive to attract hundreds or thousands of the enthusiastic people from around the world who are hungry for meaningful, exciting, healthy work they can actually get paid to perform. I think it is an exciting adventure, rescuing the future from the clutches of biospherical collapse. It's on the order of the mythical science fiction scientist saving the Earth from the relentlessly approaching asteroid and could be portrayed as crucial work to benefit country and humanity, for all the plants and animals. But the best that most aspiring ecological business people can hope to achieve today – and I see a steady stream of young people coming to my organization, Ecocity Builders in the search for the ecocity-building work I can't yet offer them – is jobs in clean up, recycling and supply of improved materials with reduced environmental impact. There are some jobs in building transit and rail equipment and bicycles and some in recycling – beyond those taken by the sad-eyed homeless recyclers with their shopping carts wandering my own town's streets. A thin scattering of architecture firms design far more ecologically appropriate buildings. But we need ecologically appropriate whole communities and an ecocity civilization economy is magnitudes removed from these good beginnings. Since our environment and our society and even our climate is showing signs of approaching severe dislocation, I believe

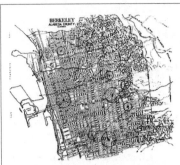

Berkeley's urban centers, present.

we urgently need a comprehensible strategy for economic transformation and these four steps are essential: l.) ecocity mapping, 2.) drawing up the list of technologies, businesses and jobs, 3.) writing the incentives, and 4.) recruiting the new pioneer communities' builder/ residents.  It all begins with the ecocity mapping based on understanding the anatomy of the healthy city.

5 - 15 years hence

15 - 50 years hence

Richard Register

Map 2. ECOCITY ZONING GUIDE—
SELECTING CENTERS, MAPPING ECOCITY ZONES

idealized center of town          •      idealized sub-centers      ▲
actual town center                •      actual sub-centers         ▲
compromise town center            ●      compromise sub-centers     ▲
existing neighborhood centers     ■

25 - 90 years hence

Urban centers positioned on a map of a natural landscape – Berkeley, California in this case – showing ridge tops, creek systems and bay shore.  The concentric circles represent rising density and diversity (or complexity of "uses") toward the centers. Moving outward from the centers, we come to areas of high priority for withdrawing from existing development, that is, removing buildings, parking lots and automobile streets as they weather and age, rather than replacing them.  In the zones farthest from the centers these things should be replaced with restored natural features and agriculture.

40 - 125 years hence

# Emerging Trends from the Global Village to the Ecovillage

## MARIAN ZEITLIN

If and when we achieve sustainable development, we will be living in ecocity communities and ecovillages. Ecocity communities and ecovillages we may call the end point of sustainable development.

In this talk I'm going to speak about four diverse trends that lead directly towards the end point, in some sense bypassing the societal. The translators have asked me to talk very slowly, so in twenty minutes I will be guilty of over-simplifying some very complicated ideas. The first principle here which is not one of the trends, but is a principle, is that a sustainable planet is a self-regulating system, just as the natural cycles are self-regulating – the carbon cycle, the water cycle, the ecological cycles are self-regulating. A sustainable planet is a self-regulating system. And we are working therefore to restore that self-regulating quality. Now one of the characteristics of self-regulating systems is that they cooperate hierarchically. Which means that cooperation at the grassroots, at the ecovillage level is an organic part – is like an organ – an organic part of the cooperation at the global level. And because this is so, trends that go directly to the grass roots level in fact simultaneously influence the global level. So one thing that we may want to do is to rethink the green movement motto: "think globally, act locally." In fact acting locally *is* acting globally because of the hierarchical coordination of self-regulating systems.

There are four trends that I want to briefly mention. The first is technological. We are now moving into what some people call an information economy and some people call a knowledge economy. And the economic shift that's taking place is a force that is taking us to the grassroots ecovillage directly. This force operates in many ways through telecommuting. In the knowledge economy, the capital for production is here in each one of

*Marian Zeitlin, Director of the Program on Agriculture and the Environment at Tufts University, Medford, Massachusetts, has spent many years in Africa working as a nutritionist and has recently been assisting traditional villages in acquiring advanced communications technology.*

us. It's knowledge. The individuals carry capital. Individuals increasingly are, in addition to banks, where the real capital is kept.

Well when the capital is here, the work can take place anywhere. And with electronic communications cutting costs, gradually there is a scaling back towards the periphery and every time there are cutbacks, there are fewer resources in the offices and there is a movement towards the periphery. Taken to its logical conclusion – and many people are already doing that – it creates ecological neighborhoods because people who live and work at home, want to have the benefits of a neighborhood and a full life. They are home more. They are naturally there for their environment. So that is one movement in very simple terms.

The second movement has to do with the globalization of corporate control and the globalization of public governance. At the present time there are 47 corporations, each one of which is more wealthy than 80% of the world's governments. And their interconnectedness is such that it is leading to a dialectic. On the one hand there is what's called the race to the bottom as people go increasingly for cheap labor at any location and for places where they can dump toxic waste materials. These companies put pressure on countries. The GATT and World Trade Organization pressure countries to lower their environmental standards for profit.

The other side of this dialectic is an ecological movement within industry called industrial ecology, the ecology of commerce and to give just a taste of what that might look like in a twenty minute presentation, one of the end points for that movement is to insist that all large appliances such as cars and refrigerators are essentially leased, that we don't own them, that the company that manufactures them has responsibility for them from the beginning to the end. When we are finished with them, the company has the responsibility to take them back and recycle or dispose of them.

So there are shifts within commerce that could have very major effects because if companies had to do this, everything would change. The whole economic dynamic would change. This is just one taste of a large movement and the dialectic between the downward spiral and the upward spiral within the global economy.

The next major movement I wish to mention is the global village ideal as it emerges from the electronic media. The electronic media are politically controlled so that Africa, for example, gets very little coverage. However the electronic media are also commercially controlled. And they must sell. To be profitable they have to sell everywhere and to be well regarded they sell green. They sell according to an ideal that has emerged as a vision of the Global Village.

I'm just going to give one example here, from an American Telephone and Telegraph, AT&T, advertisement. This ad, in Scientific American, showed four Chinese gentlemen in long silk robes out on some grass and the text reads, "there is only one human race, yet isn't it remarkable how so many things can actually separate us... like time and distance and language. Imagine if you will, a world without limits." Their marketing incentive is to get you pick up a telephone and call your grandmother or your agent or whoever. But their rhetoric creates the global village.

One other way: I think everyone here watches CNN from time to time. The imperatives for sales for CNN is equal treatment for everyone from every country, in the sense that on CNN all religions are equal. A Jewish rabbi, a Santeria priest, a Sufi mullah, a Catholic priest – they all get parallel and equal treatment. The sales dictate an ideal of cultural diversity and global equality. So that is another movement that is leading to a Global Village ideal. The dialectic of the Global Village is to create a desire for the rural village. Global village imagery is so often real ethnic village imagery. If you look at the imagery the films of some of the advertisements, we are back at the ecovillage. *That's* what they are selling.

To my last point now, and that point is post modern philosophy and spirituality. One of the trends in post modern philosophy is in response to an overflow of

imagery – we can take it back to the last point I made. When CNN presents the Pope and the Santeria priest and the Imam and the rabbi together, we are flooded with imagery because they are presented as of equivalent value. One of the results of that flood is to drive us to the particular, to deconstruct the general to the particular. In the move to the particular, we are back in Yoff, we are back considering Mame Ndaire, the spirit of the place. We're back relating to time now, to space/time coordinates at which we are because that's where our particular reality is created. We are here, we are now, that is the reality we experience. These other generalized systems are now floating all around us, so, when we are back at our space/time coordinates of where we are now, it's also a very short step to deep ecology, to viewing our space/time coordinates as part of the global space/time coordinates for all beings, and a piece of the particular, the drive for the particular brings us here, from the outside here to Yoff in a very personal way.

So those are four trends in very different areas that move us to the end point without necessarily passing through all higher policy levels. That's what I call emerging trends of the Global Village to the Ecovillage.

Yoff, the local village, prepares for Ecocity Three, a meeting of the international community – conference sign painting in Yoff, Senegal.

Women grinding millet in the traditional manner in Senegal.

*The most effective means of increasing everyone's
well-being is to invest in women.*

Robin Standish

# 4. Women's Issues are World Issues are Ecological Issues

We present women's issues here between the chapter on theory and the chapter on the history and details of African villages. Women's issues are everyone's issues and a major, perhaps *the* major link to a healthy future, so basic is the relationship between the sexes, so crucial is the nurturing role that seems to be so much in the nature of the feminine. In countless cultures around this Earth, women are the care-givers, the home-makers. Therefore, if we need to take care of the Earth and make homes of our cities and villages – get out of their way!

Robin Standish leads off with the assertion that "the most effective means of increasing everyone's well-being is to invest in women." Fatoumata Lelenta describes the social dis-investment that has resulted from Western economics, media and values. These were originally forced upon Africans against their will, eventually resulting in breaking village and family bonds and causing such knotty problems as prostitution in adolescent and young women. Marion Pratt examines a particular area of economic and social change in African fishing economies, as women gain a measure of indepen-dence and control over their economic lives – and as a result, plow headlong into major social conflicts, especially with their husbands. Then Moussa Bakayoko describes the role of the tontines, small scale savings associations, that help where banks and govern-ments dare not tread, where personal trust and the more unquantifiable values of life meet the necessity to have some cash: in ceremonies of birth, death, rites of passage and spiritual festivals, as well as in setting up businesses. The African value of "solidar-ity" – a term we heard over and over in Yoff – is partially supported by tontines, and tontines in turn, founded on this core value. "It takes a village...."

Delve deep enough and there are always more questions than answers. Reading between Marion Pratt's lines we see that the very basis of women's opportunity in the fishing economy of Lake Victoria was predicated upon the introduction of perch, which are driving out native species, and upon elimination of dangerous, if native, animals. Does this mean women's issues are, at least there, antagonistic to biodiversity? Just what is the appropriate balance between tradition and the advantages of more recently perceived rights of women and benefits – or detriments – of new technologies?

# The Most Effective Path to Sustainable Community Well-Being: Investing in Women

## ROBIN STANDISH

At least 30% of the world's people born in the 1990s will face the certainty of a grim future; their economic opportunities will be outstripped by population growth. Not only individuals, but communities and the struggle for sustainable well-being itself will be severely affected.

How can so overwhelming a challenge be addressed? What are the key leverage points for productive change? Researchers in universities, think tanks, and the World Bank have been asking those questions, and their answers may surprise some people: the most effective means of increasing *everyone's* well-being is to invest in women – in particular, in the education of girls and in women's NGOs (non-governmental organizations or non-profits). Gains from these investments are documented in areas as diverse as community health and economic productivity.

Education is associated with the acquisition of a range of skills and attributes, including adaptability and flexibility, a sense of agency (seeing oneself as a decision maker, an actor), the ability to make informed decisions, and an expanded sense of identity from villager to regional citizen to world citizen. Literacy and numeracy connect directly with economic improvement, and also increase the likelihood of active participation in NGOs where leadership and organizing skills are acquired.

Each of these skills is especially important to any people who are laboring under multiple demands, are marginalized by social institutions, or are caught at the vortex of social and economic change. Women are disproportionately represented in all of these categories. In addition, two-thirds of the world's most destitute people are women. But despite these handicaps, women are a significant majority of those who use newly acquired skills for the betterment of their families and the community. Benefits are most often demonstrated in the areas of health and economic advances including education for children, environmental protection and fertility reduction. Women are disproportionately represented among those actively working specifically for the kinds of NGOs that contribute to social change and to sustainable community well-being.

Public health benefits associated with extending girls' education are well documented. With education, women live longer and the number of maternal deaths and the number of children who die before they are one year old decline. Perhaps the most important known community benefit of increased female education is a reduction in the fertility rate. That linkage is now established world-wide, with the addition of research in Sub-Saharan Africa. Those results are further improved when men receive training in contraceptive use and parenting costs.

The value of women's time spent in economic activities is increased by education. Their labor productivity and wages are increased, with a consequent rise in household incomes, and poverty is reduced. The Grameen Bank and other financial institutions have found that women are better credit risks than men. Women who are literate and numerate have a better chance of receiving loans and starting businesses and succeeding in them.

Education is the best documented and best known means for women to acquire the attributes that they use for community well-being, but it is not the only way. Amartya Sen pointed out in her 1994 Harvard University Press book, POPULATION POLICIES RECONSIDERED – HEALTH, EMPOWERMENT AND RIGHTS, that any experience leading to women's self-efficacy will suffice, whether it is brought about by improved health opportunities, the chance to start a micro-economic enterprise, an NGO organizing and leadership training experience, or increased education. What counts is any positive experience that develops a women's sense of agency, her sense of herself as an adult person, a decision maker capable of making informed decisions – about her fertility or any other issue required of a full participant in society.

*Excerpted from a paper, talk and subsequent question and answer period. Ms. Standish is Founder and Director of the Resource Exchange, an organization that links people with expertise on specific cultures and local needs to people who can respond to requests for technical assistance.*

Karil Daniels

Robin Standish

NGOs are women's most readily observed and effective mechanisms for bringing about positive community change. They deserve attention. From well-known women's environmental projects in India to rural mobile libraries in Zimbabwe, the results of women-generated NGOs is well-documented, and perhaps the best place to explore the sources is the Global Fund for Women in Palo Alto, California. This organization funds grantees who focus their efforts on the root causes of problems. The Fund's areas of interest include female human rights, women's access to communications technology, and economic autonomy of women. Their grantees span a range of NGOs world-wide working directly on projects supporting sustainable community well-being. The Federation of African Women Educationalists (FAWE) deserves special mention for its impact on community well-being in Sub-Saharan Africa.

While any one NGO can do impressive work, their impact increases when they connect with one another. A good example of such a program is ASK, in Bangladesh, an NGO linking women living under Muslim law internationally. ASK researches ways that repressive local customs differ from Koranic prescriptions, and makes the findings known. Devout Muslim women then have Koranic support when they seek health services and education for their daughters as well as their sons, or when they organize to eliminate violence in the home.

Sustainable community well-being received a tremendous boost in 1995 at the NGO Forum of the United Nations Conference on Women in China. Collaboration between the North and the South was evident at the NGO Forum, thanks in large part to the leadership of women from the South. Evidence of women taking responsibility for a sustainable future is apparent from examining the Forum's workshop catalog. It lists hundreds of topics that link improvements for women with an advance in over-all community well-being. Often that linkage

happened, even when the workshop title did not imply it. The "Religion as Empowerment" workshop participants, for example, identified a shared vision of human rights and sustainability across all faith traditions. They committed themselves to building international networks supporting shared principles of community well-being. Educating fundamentalist extremists within their own groups and working to thwart extremists' negative impact on community well-being were additional commitments.

Because of the work of the NGOs, the UN now has a serious platform of action. Maram Rita, reporting back to Boalt Law School after the Beijing UN Women's Conference said that "the role NGOs play in fact-finding, analysis, reports, and draft interventions has not only come to be accepted and incorporated over the past twenty years, but is acknowledged as an, if not *the*, essential ingredient in the success of UN conferences on topics ranging from the rights of the child to human rights, population, environment, and development issues."

NGO preparations for the UN Women's Forum in 1995 were enhanced by the growing development of information technology. Thanks to new information technology, the cost of international NGO networking is decreasing and at the same time the speed and frequency of contacts is increasing. Cyberspace offers tremendous

Susan Felter

Vegetable vendors in the busy market at Kaolak, Senegal.

potential for increased NGO effectiveness not only in UN work, but generally.

This brings us to a more general question. How might girls' education everywhere in the south be accelerated, without the overwhelming expense and complications of conventional school systems? It is estimated that to bring Sub-Saharan educational levels up to those required for sustainable development, it would take from 10 to 100 times the present education budgets in most of these countries. Such increased spending is unlikely, to say the least. But learning with interactive media may provide an answer.

Obstacles to such an approach are considerable – maintenance and security problems, resistance to girls education from village elders and from cultural patterns, relatively high initial equipment costs – but they are diminishing and can be addressed. For example, prices for equipment are coming down quickly and Africans are used to tribal sharing of resources. What would be obviously beyond the reach of an individual villager might be accessible to a whole village or a well-organized NGO – and would tend to be used well, for the benefit of the group.

Additional concerns must be acknowledged about the use of cyberspace to reinforce existing power relationships, further widening the gap between rich and poor. Yet information technology also introduces a range of powerful tools having the potential to substantially advance sustainable community well-being. Once over the hump of original acquisition and equipment set up, communications by e-mail and other means are practically free.

Local people with animation, film, literacy, critical thinking and other skills are needed to collaborate with software design people to create relevant and effective interactive educational programs. Materials infused with local motifs and built around local concerns must be developed by and for the local people. For example, since 80% of the food in Africa is produced by women, interactive learning games could be developed for crop produc-tion. Instead of the standard good guys/bad guys motifs, Sub-Saharan African girls could learn numeracy and literacy in their own local languages by playing computer games about outwitting insects, viruses and drought. While learning to read, they would also be acquiring strategic information for optimal crop growth. They could also gain collaborative skills essential for success.

NGO projects already in operation internationally open the way for interactive video learning outside of conventional schools. A drop-in school for working children in Dakar, a video workshop for women in Delhi, and a wide range of programs including retraining of sex workers would, with minor adaptations, include interactive media learning in their programs. An international network of people interested in taking advantage of recent breakthroughs in interactive media learning is beginning to come together. My own organization, the Resource Exchange, is undertaking an extensive feasibility study in Sub-Saharan Africa beginning later this month, January, 1996. Local film makers, innovative educators and others who would like to explore possibilities for collaboration are encouraged to contact us immediately.

What is needed for information technology's positive contribution to sustainable well-being in addition to skilled technicians, teachers and interested local participants? Sufficient funding. A new generation of imaginative Carnegies and Mellons must come forward. Collaborative funding leadership from Silicon Valley's multi-millionaires could provide the needed jump-start for socially useful information technology in Sub-Saharan Africa and elsewhere.

Now to answer some of your questions.... About women's rights as human rights I couldn't agree more. I went to the NGO Forum of the United Nations Women's Conference to learn, especially from leaders from the South. But one of the things that struck me in the diversity of voices of women from the South was the unity around the issues of women's rights (which has been raised here in Yoff too). And that surprised me frequently. I would not have expected some of these

Karil Daniels

Conference hostesses in the lobby of the Yoff-Dakar Conference Center.

issues to have been raised. And they were raised frequently, vehemently and with unity from the women from the South.

Someone raised the issue of tradition. I think we have heard a lot at this conference and many other places about what looks to me like a romanticizing of either technology or tradition. And I would like for us to step back from a romanticizing about either of those things. Technology and tradition are both like a knife that can cut either way. Approaching the turn of the century, it's a good time to take stock of what parts of tradition are nourishing our souls and what parts are really holding us back and making us unhealthy, and what parts of technology are useful and what parts are destructive, what parts we do not want to get involved in. These are choices that all people have to make for themselves.

(*Ms. Standish suggested several ways in which interested people could help "maximize efforts for sustainable community well-being, assist in the education of girls in developing countries and leverage women's NGOs capacity to use information technology effectively." Here are some steps that people can take:*)

1. Connect people with information technology skills and community workers. Introduce software designers, critical thinking teachers, and community workers to one another, particularly within developing countries. Stimulate their interest in collaborating on educational software in market languages with a focus on accelerating the education of girls and sustainable community well-

being. Suggest that they contact the Resource Exchange for help with funding sources and grant proposals.

2. Encourage your country's support of the United Nations. Get a copy of the UNITED NATIONS REFERENCE GUIDE TO WOMEN'S HUMAN RIGHTS, and check the chart in the back to see where your country stands. Encourage support to ratify UN agreements your country has not yet signed or is not enforcing.

3. Get Janice Jiggins' book CHANGING THE BOUNDARIES – WOMEN-CENTERED PERSPECTIVES ON POPULATION AND THE ENVIRONMENT. Refer to it as you make decisions, whether you are a planner, policy maker or community worker, and encourage any receptive colleagues you know to read it. Make copies of Jiggins' ten point "Agenda for Action: What to Do" and distribute them to interested people.

4. Support the Global Fund for Women with your contributions. Write to them and get their information to share with your friends. Get friends to send them a donation in lieu of a birthday, Christmas or other present –as a step toward sustainable community, this has the additional benefit of modestly reducing material consumption.

# Causes and Prevention of Prostitution in in Young Women

## FATOUMATA LELENTA

Fatoumata Lelenta

In pre-colonial Africa, the house was inhabited by large families comprising three or four generations, with responsibility for family well-being belonging to the oldest male. The household was the social unit of production, and the resulting interrelationships gave a far-reaching strength to the family. Today, profound and rapid social transformation is eroding the family's traditional control over its members who increasingly are engaged in external activities. This change has deprived families of economic and moral support once offered by their larger community. Also, children are no longer effectively joined to a group to teach them the concepts of good and evil and guide them through their childhood.

Formerly, the traditional social organization was rooted in the village, symbolizing fixation to soil of the lineage group which includes all of the descendants of a common mythic or mythicized ancestor. In addition, the eldest member of the village had the moral authority role.

Traditionally, the upbringing of children was tackled by all of the villagers. At the village level, anyone was more or less parent, friend, sibling, cousin, knowing each other and living together with little difficulty. Without the village's social control over children's behavior, juvenile delinquency is dealt with as violation of law, exposing the child to penal sanction. Under-age prostitution is one such delinquency when the violation involves a young woman of 12 to 21 years of age. Why do these young women engage in prostitution? The root causes for maladjusted young women begin at a young age, and they must be addressed at family, social and economic levels.

The family plays an essential role in the shaping of the individual during infancy. It is in this environment that the child needs to have basic needs generally satisfied (food, clothing, affection and security). If this does not happen, the child looks elsewhere for satisfaction. In the context of today's social transformation, the family environment tends to lose some of its basic functions.

*Fatoumata Lelenta is a specialist with the Palace of Justice, Dakar, Senegal.*

Economics plays an important role: the lack of job opportunities leads to vagrancy and prostitution among young women upon reaching puberty. An investigation carried out among young women engaging in prostitution in the Dakar-Plateau showed that most of them frequent soldiers from the French naval base. They earn much more with them (from 40 to 60 dollars a night) than they could earn otherwise.

The prostitution of young women is one of the most painful forms of social decay, because women have a special status as wife, creator and nurturer. Girls from age three on must be guided toward different and better behaviors. It is necessary to point out the kinds of work which are suitable for them.

What incentives can be given to these young women to encourage and enable them to meet their needs in other, healthier ways? Any prevention, detection or treatment program must address a range of causes and factors, including inadequate socialization leading to submission to men and young men, lack of accurate information about sex and lack of dialogue between parents and children.

Putting an end to this scourge will require comprehensive and cooperative efforts between: parents, families, local communities, teachers, decision makers and planners, funding bodies and agencies, the media, and, young women themselves. All parties must understand what prostitution offers, and we must continue raising awareness of prostitution's consequences. We need initiatives by decision makers which take affirmative actions for young women by developing employment in appropriate sectors (e.g. fabric dyeing, embroidery, dressmaking). Conditions can be improved for young women through education and literacy campaigns. We must be able to provide our children with all the time requisite for structuring their personalities, since they remain the most precious gifts with which God gratifies us.

# Income Generation and Gender Conflict in Fishing Economies in East and West Africa

## MARION PRATT

The long-held perception that women play an insignificant role in fishing economies is a direct result of researchers' tendency to define a fishery uniquely in terms of the capture of fish and thus to direct their attentions and questions to men. It is true that in only a handful of fishing societies around the world do women work side by side with men to catch fish. The reasons for women's limited involvement in the capture of fish are many and varied. The dangerous, high risk nature of fishing has fostered certain associated ritual beliefs that have served to restrict women from participation in fisheries activities. But the most common reasons given for women's limited participation appear to be identical to those given to explain women's limited participation in any other activity that is customarily male-dominated, those being: women's primary attention to time-consuming reproductive and domestic tasks reduces their opportunities for travel and availability, cultural proscriptions inhibit the contact of women with unrelated men, and women's limited access to training and start-up capital restricts their ability to undertake income generating activities. However, women in fishing communities become eminently visible when the definition of a fishery is expanded, correctly, to include activities that are equally important to the success and efficiency of the fishery as a whole.

In addition to harvesting, all fisheries systems have two other crucial sectors: processing and marketing. Due to the high perishability of fish and fish products, all three sectors are inextricably interrelated and interdependent. Tied importantly to fish harvesting, processing,

*Marion Pratt works for the United States Agency for Intentional Development, formerly with the Agriculture and Food Security Office and presently with the Foreign Disaster Assistance Office. Her plans to attend the conference, like Rusong Wang's, were derailed by the US government budget impasse, and in addition, stymied by the worst snow storm to hit Washington, D.C. in 500 years. This is a lightly edited version of the paper she intended to deliver in Yoff.*

and marketing are also ancillary activities in which women often play a dominant role. These include net making and mending, fuelwood collection, beer brewing, preparation of lodging and meals, and the provision of informal credit and sexual services.

In a few extensively documented cases – most notably in coastal communities in West African countries – men and women fisher folk work together very closely, providing examples of efficient, productive cooperation. In these instances, wives buy fish from their husbands to process and sell. Sometimes these same women extend credit to fishermen from their savings.

Women in East Africa have only more recently become as importantly involved in income-generating activities within fishing economies as West African women have been. Let us look at one example, in northwestern Tanzania, on the shores of Lake Victoria.

The peoples of Kagera Region in northwest Tanzania belong to highly stratified, patriarchal societies that in pre-colonial times were ruled by kings. The kings maintained their power by means of control over a feudal agropastoral system based on plantains and cattle. Ideally, women in these societies remained at home, caring for the family plantations and the children. Today, of the economic activities in Kagera Region, the local fishery industry represents the most potential and promise for Haya people, due to rather unusual and somewhat fortuitous circumstances. In the late 1950s, British colonial authorities introduced an exotic species of predatory fish, the Nile perch, into the Ugandan waters of Lake Victoria to revive the fishery. Quite unexpectedly, the perch population exploded in a dramatic fashion, and now, forty years later, it provides the basis for a thriving fish export business in Uganda, Kenya, and Tanzania.

Until the arrival of the perch, Haya and Zinza women were little involved in fisheries activities for three main reasons: 1.) there were few tilapia – the main commercial species in the lake before the perch – in the northwestern waters; 2.) war parties from kingdoms in Uganda frequently raided the western shore; and 3.) dangerous wild

Bill Mastin

A display at the Women's Museum on Gorée Island near Dakar shows various traditional baskets used by women for carrying fish.

animals inhabited the beaches and inshore waters. The elimination of inter-kingdom warfare and dangerous wild animals during the colonial period, especially hippopotamuses, which overturn boats, and crocodiles, which attack people in water and on the shore, in combination with the new opportunities in the improved fishery, encouraged the growing participation of local women in fish processing and marketing.

Though there are more women in fishing communities now than there were during pre-perch periods, they are still significantly outnumbered by men. Women face difficulties gaining access to control of benefits derived from the male-dominated fishing economy. The United Nations Development Program/Food and Agriculture Organization project, Integrated Fisheries Development in Rural Fishing Villages, Kagera Region, was initiated in April 1991 to facilitate the development of the Kagera fisheries economy. It was thought that the below average incomes of Kagera inhabitants might be increased through the vitalization of the under exploited fishery. The objectives of the project were to increase the productive and income-generating capacities of the artisan fishing people, by improving fishing and processing technologies, and reducing post-harvest losses.

Like women in West Africa, women in Kagera fishing communities face particular constraints in their struggles to provide for themselves and their families. These include limited formal education and training, and restricted access to capital and labor. In an attempt to address these problems and to encourage women to participate in income-generating fisheries activities, the Food and Agriculture Organization (FAO) project initiated a small-scale revolving loan program to help women already involved in fish processing or marketing to expand or streamline their businesses. The average amount of each loan distributed was $US 150 (1992 rate of exchange). By 1994, about 40 women received loans. The overall repayment rate was over 82 percent, a clear indication that women in this region were not only interested in income-generating activities but would use project funds responsibly.

The following case study of one of the loan recipients reveals the types of problems women face as they struggle to run businesses in competitive and sometimes hostile surroundings. Mama Mary, a Sukuma living in Biharamulo District, was married to a farmer at the time she applied for and received a loan from the FAO integrated fisheries project. Initially, she used the money in her perch-frying business, but, complaining that the constant exposure to smoke bothered her lungs, she soon invested in a small tea shop. Her husband often took the profits she made and used them to buy beer; she said he was a heavy drinker. Mary complained, "After I received the loan I was held responsible for buying everything for the household." Running the business efficiently was difficult because of her husband's constant requests for money. Why didn't she refuse to hand over her money to her husband? Mary has a long, faded scar that runs across her forehead from hairline to eyebrow. The local fisheries officer informed us that several years earlier, Mary's husband had come home drunk and hit her with a piece of firewood. Believing that the only way to continue her business successfully would be to leave, she moved out of the village and set up her shop along the road near a settlement to the northwest. She explained that it was not easy to run a business by herself, but free of her husband's disturbances, she was convinced she could continue her work successfully.

Women have on average many fewer years processing experience than men have, and, lacking their own start-up capital, are obligated to obtain it from their husbands, friends, or kin. Similarly, many women cannot afford to hire labor and so must squeeze the activities into an already full schedule of domestic and agricultural chores.

Men throughout the region are worried about women's control over money. The fishermen did not agree with extending credit to women as they feared that the women would become stubborn and would not listen to their husbands, thus leading to the breakdown of marriages. Men approved of their wives contributing to 'their' – the men's – household income, but said that the contributions should be limited so that women did not get 'above themselves' and tempt men to divorce and/or take a second more docile wife. The divorce rate among women in business is said to be high.

Men are not joking when they complain that women should not become too powerful, or be allowed to exercise unilateral control over money. The payment of bride price and the fact that many husbands supply the start-up capital for their wives' income-generating activities legitimate – from the husbands' point of view – men's access to any significant income their wives generate. Often, for this reason, women keep their incomes secret from their husbands. Nowadays, however, the material, social, and economic benefits of being a good wife or an obedient daughter are no longer guaranteed, as husbands leave more and more household responsibilities to their wives. Simultaneously, market relations and commoditization have opened up opportunities for women to acquire more cash and productive resources in their own right.

There are many lessons to be learned from the study of gender dynamics in fishing communities in East and West Africa. First, neither women's nor men's behavior and decision-making strategies analyzed separately, without reference to the other, is fully meaningful. The simultaneous study of men's and women's behaviors – and the interactions between the two – has proven to be a useful framework of analysis with respect to both past phenomena and future projections. Further examination of the psycho-social dynamics of gender interactions in fishing economies would help to reveal the price that innovators are paying for modifying their gender roles, and thereby challenging the local gender system.

Many women in African fishing communities have demonstrated their ability to transfer newly available wealth – and the authority that wealth engenders – into their own hands, and by doing so actively reject societal "belief" in the inevitability of patriarchal control. Such manipulations of circumstance, however, have exacerbated household conflicts between husbands and wives as the balance of power has shifted. Though circumstances change relationships between women and men, ideologies of gender are typically slow to follow suit, and where the chasm between the ideal and the real is widest, one often finds associated domestic turbulence in the form of argument, alcoholism, violence, and divorce.

The women interviewed for this research were unusually open about their personal circumstances because, I believe, they all felt singularly entitled to the loan money they applied for and received. The capitalist system as represented by the fisheries project, with its emphasis on individual success, for better or for worse, encouraged women to think more about their circumstances, their potential, and their lack of power in the domestic sphere. The research into the fisheries project reveals the power of capitalist relations to generate and transform social, and in this case, domestic relationships.

Currently, in many fishing economies in Africa, foreign fish traders – bigger capitalists – are buying up much of the fish for export, driving up prices and driving out small-scale competitors, such as women. It remains to be seen if, under these conditions, women will be able to keep a hand in these economies, or if the commercial fisheries themselves will remain viable for much longer as many fish stocks around the world are seriously over-exploited. Evidence from the past strongly suggests that African women will continue to find alternative strategies for survival, but perhaps at similar costs to their physical and psychological well-being.

# African Tontines

## MOUSSA BAKAYOKO

In Africa there is a well-known type of association, the *tontine*, which functions as a kind of grassroots savings and loan. Since banks do not lend money for expenditures like hospitalization of the ill, funerals, weddings or birth celebrations, citizens organize on their own without the banking system's help, forming groups for their mutual economic assistance. The tontines help their members save money and deal with large expenses when they are without savings.

In a typical tontine each member deposits a set amount of money each month and then receives the yearly total amount of her savings when her turn comes. If a member's turn comes in the first part of the year-long period, she still receives the full amount of twelve times the monthly installment. Her yearly withdrawal works as a loan. For the member who receives her payment last, the tontine is essentially a saving mechanism that year. However, the last member to receive her turn is often given the first turn the following year, giving her a much larger sum of money with which to invest. In this type of tontine there is no interest earned or charged to members. The order of turns is usually established each year by drawing lots, although sometimes turns may be bought at auction. Some associations do pay interest on savings and charge interest on loans of varying amounts.

Citizens who group together to form a tontine usually have something in common – a common trade or ancestor, civil servants working in the same office, or women living in the same compound. It is essential that the members of each association know each other well and share a strong trust. Usually a tontine is exclusively a women's group, although sometimes a men's group, and may create an important element of solidarity within the community. The number of members range from a handful to several hundred, but the preferred number is a multiple of twelve, matching the monthly periods when

---

*Based on a talk and paper by Moussa Bakayoko, a sociologist in the Urban Development Office of the Municipality of Dakar.*

each member takes advantage of the large payout. Once a month a lump sum of cash becomes available to one member (or group of members) at the time of the monthly meeting.

A tontine usually has statutes or rules and may be managed by a board consisting of a president, who takes charge of meetings, a secretary, who registers any decisions made and takes the minutes in meeting, a treasurer, who collects and distributes money and keeps the accounts, an account commissioner entrusted with controlling the treasurer, and sometimes a censor who can discipline members.

Often the annual distribution to each member is used to start up an enterprise or build a home. With some groups only part of a member's annual contribution is returned in a lump sum, and the rest is mutually invested. Sometimes schools or other community services are the goals of the mutual investment. Another facet of some tontines is called mutual aid activity: the group makes a contribution to a member in need according to its rules. For example, one Congolese association contributes 1,500 CFA in the event of the death of a member's father, mother or spouse. The contribution amount is 500 CFA in the case of the death of a member's child and 2,000 CFA in the case of the member's own death.

Other variations in investment activity might include building a hall and buying chairs and awnings, etc., for rentals, or buying a vehicle for transport between village and city. Members normally receive a discount on use of these services as well as sharing in their profits. A tontine also might invest in an enterprise like a retail shop or gas station. Some such associations pay extraordinary rates of interest on savings – up to 70% in some cases – giving excellent protection against inflation, which can reach 15% per year and higher.

These working class associations are a magnificent example of self-organization within neighborhoods. In an economic context where the banking system is inaccessible, maladjusted, and too expensive, they provide an alternative solution, easing the financial

Susan Felter

Yoff, Senegal. Tontines help start up small retail enterprises.

difficulties of their members through the pooling of resources.

Although tontines are widespread throughout Africa, they are often misunderstood or underestimated by the media and the technical services. Rarely are they mentioned in the press and the government gives them no support. In fact, since few governments officially recognize the tontines as entities, they can rarely open an account or carry out a legal action in the courts. In order to broaden the tontines' popularity and influence in the future it will be necessary to counteract this tendency through education, activism, and cooperation between the groups. Short-term goals include gaining access to technical services and networking to share advice on management and investment techniques.

Yoff, Senegal.

*How does the traditional village cope with the disorientation in cultural and social values that Dallas and Dynasty bring sizzling across recently installed electric wires to the remotest village compounds? How too, does it emerge whole and sustainable in the face of political pressures of a 'modern' Senegalese government and the environmental degradation of an encroaching city, Dakar?*

Nancy Willis, journalist writing for *Annals of the Earth*

# 5. Traditional Villages in the Dakar Region

If you missed the conference, the next best thing to being in Yoff and Dakar is reading this chapter. Here we hear from the villagers themselves about their villages. Their history, their spiritual beliefs, and their conflicts with one another and approaches to solving these conflicts are revealed in about as much depth as these few pages allow.

As in anthropological reportage when it strives for objectivity, there is little interpretation here, little reticence and no excuses for different ways of life and doing things. In fact there is pride revealed along with the many problems of a society in profound transition. Sacrificing bulls to placate village spirits? Without attempting to draw cultural equivalents in the visitor's culture, such as the bloody sacrifice of a quarter-million people a year to the god of automobility, without judgment of self or others, the first two reporters leave it up to us to try our hand at understanding the meanings embodied in the history and political structure of villages of the Dakar region, as depicted by Cheikh Tidiane Mbengue, and in religious beliefs and ceremonies, as described by Assane Sylla.

Ismaela Diagne, however, draws his own severe conclusions contrasting the almost idyllic past lives of the traditional villagers on one hand, and on the other, the current destructive and immoral onslaught – that's how he feels – of expanding big cities like Dakar and Thies. Meantime, villages are abandoned far from cities across his country. Meantime, the land of the villages is expropriated without notice, not by colonial powers but by the newly independent Senegalese government on behalf of self-seeking developers and power-loving politicians. In his talk and paper we see the legal and economic workings that are rapidly eroding almost everything he holds dear. There is theft and deception here enough for a tragic epic tale, of which this is the barest introduction. But also, in courage and optimism for the future, and with a political sense for getting things done, Diagne is just as detailed in possible solutions as he is scathing in his criticism. And you will even see ways in which people struggle against developers, governments and just plain disrespectful fellow citizens, to recycle, farm, prevent toxic dumping and re-establish healthy environments. Many clusters of concerned individuals, often women, people of small powers are struggling for solidarity – and maybe beginning to succeed. If ecocities rise from the sprawl of Dakar and the smoldering of traditional village values, they will come from seeds such as these.

# An Introduction to the Traditional Villages of Yoff, Ngor and Ouakam

## CHEIKH TIDIANE MBENGUE

Some say the word "Lebou" came from the Wolof word "lebe," which means "to tell tales or fables, to be clever or to drop hints." Another theory is that "Lebou" originated from "lubu" which means a warlike soldier, or from "lebukay-ba," which means "the spot where we can find someone to lend us some money or food in times of hardship."

Whatever the meaning of the word "Lebou," it is a fact that the Lebou are a sizable community scattered in traditional villages on the outskirts of Dakar, the capital of Senegal. They include Yoff, Ngor, Ouakam, Yen, Rugisque, Bargny, Mbao, Yeumbeul, Deen, Diander, Tor, Toubab Dialaw and Hann.

These people originated in Ethiopia and began their migration around 1,500 years ago. They lived in Egypt for some years, then migrated to Libya where they were driven out by Persian and Assyrian invaders between 616 and 667. Thence to Mauritania, where in the 10th and 13th century they were driven out by plundering Moors and then to the Senegal River Valley which they left in 1569 to avoid being sold as slaves by the Kings of then Cayor, then to the village of Diander from which some of them went on to found Lebou villages in the coastal area near Dakar. The first such village was Ngokhekh, which is now the site of the national soccer stadium. The Bén clan from Ngokhekh founded the 12 districts of Dakar and, somewhat later, the Soumbadioiunes clan founded the villages of Yoff, Ngor and Ouakam, approximately 400 years ago. To accomplish this, the Lebou expelled the Mandinkas, who had been the first settlers.

Thus the Lebou migrated from the eastern Horn of Africa to the western most part of our continent, as the history relates, because they were unwilling to yield to political domination. This adds weight to the theory that "Lebou" comes from the word "lubu."

Through colonial times they succeeded in preserving their social and political traditions around Dakar and

down the coast to the south. Said the French commander of Gorée writing to the Governor on December 15, 1864, "We pay duties for our water supply and other items in Dakar. If we were the rulers of this area it still would not have resulted in our owning land. The population has an indomitable passion for independence. We cannot carry away a sack of millet, a bundle of firewood, a pail of water, without paying another duty to the chief of the village from which we take it."

Two fundamental features characterize the social and political organization of the Lebou. On the one hand, it is a democratic, republican organization. On the other, it is an organization which fundamentally articulates the solidarity between individuals: men and women, children and adults, and between different families. It exhibits a kind of solidarity which is vertical and horizontal at the same time. It is an organization which attempts to prevent power from being concentrated in the hands of one family that would be tantamount to a monarchy. In view of the Lebou's commitment to freedom of thought and action, such a monarchic system would not last long.

The Lebou's political system is, thus, a democratic political organization which challenges the theory that the origins of democracy lie exclusively in the West. The Lebou organization features a two-headed executive power and a two-chamber legislative system.

The Djaraf is head of the community, assisted by four high officers in a hierarchical system, but a hierarchy backed by interdependence among the different leaders.

### The Village of Yoff

In the case of Yoff, the villagers set up a political organization founded on the kheets, or clans, which are matrilineal. And each clan is said to have a protecting Rab, or spirit. The chief functions, though exercised by specific kheets, are interdependent. The Djaraf, for example, is head of the village community, but he cannot make a unilateral decision committing the whole community without first taking into account the others' opinions by consulting them. His main functions are to ensure the

*Excerpted from a talk and a paper by Dakar historian Cheikh Tidiane Mbengue.*

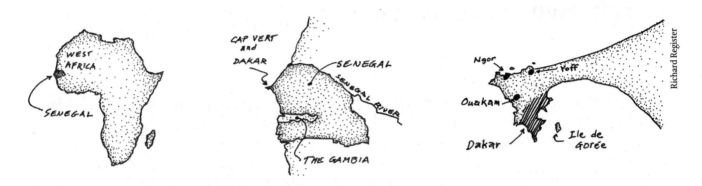

From the continental to the national to the local scale. Yoff, Ngor and Ouakam are located on the extreme western promentory of Africa: Cap Vert peninsula.

spiritual protection and economic prosperity of the village. He belongs to one of three kheets and is enthroned by the Saltiqué according to a well defined ritual where he commits himself to protect the village against any foreign invasion or misfortune.

The Saltiqué is the Minister of Defense, Agriculture, Fisheries and Worship. He is chosen from among two Kheets different from the Djaraf's, and is selected on the basis of his mystical powers and moral qualities – his mystical powers because he must be able to ensure to farmers excellent crops by causing rains, and to fisherman a good yield by placating the sea spirit Mame Ndiaré. He must be able to foretell and ward off misfortunes before they prey upon the community. He must be the guardian of the wisdom of the elders, and assistant to the Djaraf in discharge of his duties as head of the village. The people must know his past and, though he must be an obliging man, his moral qualities must be incorruptible. He is enthroned by the Djaraf during a ceremony lasting a week during which he perfects the knowledge which forms the basis for his power.

The Ndeye Dji Reew is the third major figure in this socio-political structure and is selected from a different kheet from the Djaraf's and the Saltiqué's. He plays the role of Foreign Affairs Minister. He is responsible for receiving, welcoming and directing anyone and anything coming from abroad, be it a messenger in wartime, an emissary bearing riches for the community, or an ambassador from a neighboring village. He is a mediator for the diverse Lebou federative states. A clash of competence and power may occur among the rulers; it is the Ndeye Dji Reew's job to settle things by reminding each party of

the limits of his prerogatives and by convening a meeting to reconcile them.

The Djambours are an elders council equivalent to a Senate and they are consulted by the three executive authorities mentioned above in the decision making process. They are not chosen by the executives but as representatives of the village at large. They must be at least 65 years old, and each one honest and of good morals. Each must be a third generation villager, that is, must have at least two grandparents who were born in the village. Membership does not exceed 30, but in fact, it is at times difficult to find enough interested persons who fulfill all the requirements. This council meets on Fridays inside the Great Mosque so as to examine issues that have been submitted either by executive leaders, or by groups of young men or women.

The Freeys are militia members who have been entrusted with keeping order, discipline and cohesion and also with putting into place mechanisms for solidarity through projects of common interest. They must be at least 50 years old but not more than 64. In addition to their police responsibilities, when a member of the community is kept away from his fields or fishing activity by illness, or when someone is a victim of a natural disaster, it is up to the Freeys to organize a solidarity campaign by collecting funds in order to support him, or by organizing a team to help him work his fields or repair his fishing equipment or his house. The Freeys have the power to settle conflicts between community members by inflicting penalties. They also manage the royalties paid by foreign fishing boats operating in village waters. All these funds are to be used for rehabilitation, repair and upkeep of the Great Mosque, the cemetery, and the funding of religious and traditional ceremonies such as the Tuur, which is taking place at the same time as this conference, the Third International Ecocity Conference.

## The Village of Ngor

When the first people of Ngor arrived at the extreme western end of the Cap Vert peninsula ("Green Cape" in French), the westernmost land of all Africa, they found a large expanse of trees they cleared for agriculture over a period of many years. The newcomers showed courage in clearing lands and honesty in dealing with others in the region. Thus they won the name "Gor" which in Wolof means honesty, dignity and pride, that through spelling alteration over the centuries became Ngor. The meaning holds today, for the inhabitants have remained true Lebous in spite of the tourist boom in their village.

With respect to political and social institutions prevailing in other Lebou villages in the Cap Vert region, Ngor stands out as a special case due to geography. Ngor's boundaries are fixed by the villages of Yoff and Ouakam and by the Atlantic Ocean. It is a village that has withdrawn from its agricultural land into the western rocky area that is not adequate for farming; the Dakar International Airport covers much of its former lands, as well as much of Yoff's. The absence of large scale agricultural activity has influenced the political organization insofar as Ngor doesn't need a Saltiqué. It is unique in its identity as a village specializing in canoe fishing and tourism. Its population is extremely proud and has always had a sound relationship with its Yoff and Ouakam relatives. It has never felt a need to elect a Ndey Dji Reew to settle disputes which might erupt between Ngor and its Lebou neighbors.

## Religious Organization

Among the Lebou villages we are dealing with a real mixture comprised of three main elements: traditional religious life that remains in the hands of the individual, the magical realm, linked to beliefs held by many men and women and Islam, the inspired religious life which is chiefly the concern of men.

People tend to define the Lebou as polytheistic based on various cultural practices that celebrate spirits that the uninitiated regard as gods. But one needs to distinguish between good and bad spirits and between believers and those who are not. In addition, the Lebou of the Cap Vert region give highest standing to Muslim spirits, as shown by the incantations performed during rites. The first words of such incantations state the supremacy of God the creator, the beginning and end of everything. "God is the Lord of Lords, the King of creation, and Mecca is the Queen of all holy cities."

These opening formulas give evidence of all the precautions taken so as not to regard any spiritual shrine or place as equivalent to Mecca, any rab (spirit or its totem) as a higher power who can intercede to accomplish the will of God: creator of heaven, earth, human beings and genies.

The Lebou practice the ritual of Ndeup for the sole purpose of making a connection with the creatures most strongly connected to the ancestors by lineage. Each Lebou family which has immigrated to Cap Vert has its own genie which in turn depends on the genie who watches over the entire community. The genies of each community are the following: Leuk Daour for Dakar, Mame Ndiaré for Yoff and Gorgui Bassé for Ngor.

One placates the rabs (totems) by sacrificing oxen, goats and chickens whose blood is poured on the totem's altar with millet paste and pure curdled milk. Afterwards, the sacrificed animal is distributed to the whole family or community.

Among the Lebou magic is for protection, healing and success. Indeed, the Lebou, like other Africans, is a person with fears. He must not only placate and establish relationships with the God of Islam and traditional ones, but he must also protect himself physically, visibly and materially against danger. He needs protection against sorcerers and diseases.

Thus, for example, the Lebou distinguishes and has a talisman for stomach and headache ("tere"), one worn on

Bill Mastin

The Ecocity 3 Conference coincided with the annual "Spirit of the Village" festival, shown here in one of Yoff's narrow streets.

the side against malevolent spirits or snake bites ("tere wet"), and one worn belt-like on the waist against mortal dangers or to ensure to husbands the fidelity of their wives ("ndombos"). Houses, canoes, dumping grounds ("seunes") are also protected against bad influences.

Islam was not fully integrated into Cap Vert society until the second half of the nineteenth century. It was in the Lebou community that Mahdi Seydina Limamou Laye aimed his appeal leading the Lebou to practices which plunged them into the light of Islam.

Taking stock of these beliefs and religious practices, what is striking is the division between women's religion and men's religion. This situation is of great interest. In fact, the complex mechanism of conservatism assures the Lebou of a sort of balance of protection and of permanent power.

### Traditional Ways of Life

One of the particularities of the traditional Lebou way of life is the great influence of the women. Economically independent, the Lebou woman manages her business successfully by reselling farm or fish products she buys, sometimes from her husband. Many women's

inheritances are handed down to their children under the control of the maternal uncle without the husband's interference.

Despite proximity to Dakar, the Lebous remain cultivators and fishermen faithful to their family traditions, fishing in the Atlantic and cultivating crops in gardens surrounding the villages and, farther away and cultivated only in the rainy season, a belt of millet and peanut that alternates with fallow land. However, the land registration system in use since 1930 and the speculation which followed has swallowed up most of the fields in favor of houses.

For foreigners, the social cohesion of the Lebous is probably most evident in fishing practices, but stronger still, evident only within the family compound, are the ties among people having a common ancestor. Within this world, the entire family and the house, or compound in which they live appears fixed. But in fact, the house (or "Keur") is falling apart.

How we prevent that should be one of the most important questions faced in this conference.

# Traditional Beliefs and Local Government Systems

## ASSANE SYLLA

A people's culture is its entire set of discoveries and scientific technological innovations, its ideological, moral and philosophical choices, its religious beliefs, its artistic creations, its social and judicial institutions... its values which ensure its security, its social equilibrium, its spiritual flourishing, its well-being, and its happiness.

In addition, it is the achievement of a long accumulation of successes throughout its history, each generation passing on to the next an inherited and improved legacy, at the cost of sustained efforts and sometimes painful sacrifices.

One understands easily why there is not a homogeneous culture for all peoples: the differences in history, geography, and climate result in the diversity of recorded cultures, a priceless treasure which benefits us all. Furthermore, the mixing of peoples engenders a greater awareness of universal values, thus providing more security, solidarity, love and peace to humanity.

In order to characterize the people's culture of this western region of Senegal, a people who traditionally make their living from the area, I would like to say that it is founded in humanistic values such as the honor, the respect for each person and love based not in theoretical rationality, but rather in beliefs elevated to a sacred level. In this region the sacred occupies an important place at the psycho-social level and political-administrative level as well as the level of religious belief and practice.

Among people everywhere there is a psychic élan which implies the idea of excellence or of the superiority for feelings of respect, love and fear. Among human beings this appears to be universal, and along with it comes the execution of a rite of allegiance or the adoption of relevant recommendations.

Here in the westernmost region of Senegal, more than in most places, the sacred is presented as a strong idea, a non-disputable belief, and a guiding principle of economic activities. Here the sacred character of a being

*From a paper by Assane Sylla, researcher for the Institute of African Studies, Dakar.*

or of an institution is still under the watch of supernatural beings ("God" for the monotheists).

Lebou traditional beliefs reveal the existence of spirits called "rab" or "tuur" intimately associated with the activities of individuals and groups. The ancestors have identified themselves for centuries with these spirits.

A number of families have in their back courtyards an "autel" where rituals are directed at the tuur, a protector spirit. Each village – Yoff, Ouakam, Rufisque, Dakar – has a tuur which periodically requires ceremonies or allegiance rituals like those that we observed yesterday in Yoff for Meme Ndiaré.

Similarly, the tuur delivers useful information through an intermediary, intervening to cure a sick person, or help an undertaking toward success. It also predicts events, grants a favor or helps deliver a victory against an adversary.

These traditional beliefs play, among other roles, that of being guardians to the social order and moral values. Every citizen knows or believes that the tuur of the family or of the village can inflict a painful punishment on those who violate the sacred space: illness, fire, failed enterprise, bad crops, etc. The citizen who thinks she or he has fallen victim to these punishments has to resort to ceremonies of exorcism, of which one phase consists of sacrificing an animal (chicken, goat, ox) on the altar of the tuur as reparation for the mistake and to appease the tuur's anger. These beliefs are fought by orthodox Islam, including one of the most eminent chiefs of Yoff. Yet the rituals persist as a relic of an old paganism and permeate the life and the activities of a great number of Muslim citizens.

Traditional beliefs have also had real impact on the political-administrative organization of the society. The election, enthronement and the career of the traditional rulers do not escape these beliefs.

For example, before the election of the Djaraf, the Ndey Dji Reew, the Saltiqué, the Djamburs and the Freys, consultations are led by rulers responsible for the choice of candidates, but diviners are also consulted to in turn

Susan Felter

The elders of Ngor and conferees meet next to the mosque under the limbs of a 600 year-old tree.

solicit the opinion of the tuur. After the election, the enthronement is preceded by a retreat intended to initiate the elected leader into the mystical knowledge. The support given by the tuur constitutes additional sacredness for the newly elected leader which strengthens his power.

Thus, the system of government which has been instituted in each village for centuries has benefited from great stability thanks to the sacred in both traditional beliefs and Islamic orthodoxy. It has also benefited from the balance of powers conferred on the diverse elders and also from a decentralization which makes each village administratively autonomous. Indeed, one might well fear that a secularization of institutions and of the

hierarchy of social relationships might cause a dangerous drift leading to immorality and anarchy.

# The Traditional Village Under Assault

## ISMAELA DIAGNE

I would like to begin my talk by thanking the organizers of this important international conference: they give me an opportunity to discuss some problems which arise from the fact that Ouakam, one of the largest villages of Senegal, is located in the suburban zone close to Dakar. This capital city grows like a cancerous tumor, creating serious social dysfunction, ominous economic imbalances, heart-breaking cultural conflicts – along with all the other nuisances of consumer society. When I describe the situation of Ouakam, be aware that I speak also of Ngor, Yoff, Yeumbeul and other villages on the fringe of Dakar, and the 17 villages such as Grand Thialy and Selma, that are threatened to be swallowed by the expansion of the city of Thies. At the same time, the villages of Ngerane, Tainabe, Missira Keur Mbaye and others have been deserted by their inhabitants, leaving behind them arid fields and dry pastures to swarm around the city of Touba, in hope of a better tomorrow.

Beyond the specific situation on which I report, it would be necessary to see, or even better to feel, the many aggressions victimizing these villages located close to spreading cities the development of which creates always serious problems that are at the center of the preoccupations of this conference.

### Traditional Villages of Yesteryear

To a large extent, in the traditional society, the nature and organization of communal activities depended upon the use of space. A private space was at the disposal of each family. The "wanack" was a space where women performed domestic chores, took care of the children, gave them the first elements of their education with the help of the grandmothers, guardians of the tradition. In the wanack, the residents of the same compound might meet during their leisure time in the central yard – the "ett." There, children being brought into social interac-

Mosque at Touba, Senegal.

*Ismaela Diagne is a professor of French and school director.*

tions for the first time in their lives were at liberty to play freely under discreet but effective supervision of all family members. When children were not under the immediate control of their biological parents, they became the collective responsibility of the extended family. Adults talked quietly, exchanged information, and availed themselves of the skills of the barbers. The importance of the central yard became obvious during the celebration of family events, (baptisms, weddings, funerals, tatooings, circumcisions ) particularly at the "Tamkharit," the end of the lunar year when the extended family met, shared a meal and prayed together.

Sheep and hens could roam around from unit to unit without any risk of being stolen, individual property being sacred to all. Women grew staple crops in the immediate vicinity of the tribal land in order to mix work in the fields with demanding, domestic chores. In these settlements, there was no place for idleness, therefore for envy, slander, or back-biting. The days were filled to everyone's satisfaction.

If family conflicts arose, conflicts seldom went beyond the family circle or the neighborhood. People took complete responsibility for their lives, tactfully, without complacency, with a discretion taught by experience and a will to contribute efficiently to the unity and the cohesion of the basic cell to which one belonged. When conflicts reached a certain pitch in spite of several mediations, the file was sent to the "pinthie," the neighborhood mosque. There the Diambours settled the matter, spelling out the remedies to the injustice that took place, and setting the stage to bring about the reconciliation, which they would administer.

The pinthie is not only a place of worship, but also a place for conviviality, solidarity, relaxation and meditation. That is why, even today, in many villages where traditions are still observed, it is at the pinthie that deaths, baptisms and weddings are proclaimed, that vaccinations and school registrations take place.

Under the "Mb'r," a roof suspended on pillars in front of the mosque, the elders sit in shade and recall the past. Wisdom is proclaimed there and, sometimes, the next generation, sitting a bit away in sign of respect, is addressed with pride or worry. Farm workers, coming back from the fields, place at the feet of the elders, mangoes, watermelons or beans. Sometimes the elders take a walk along the seashore, where the fishermen give them fish as a token of their moral obligation toward the seniors of the neighborhood.

The heads of the district and the Imam of the mosque celebrate all kinds of ceremonies in the pinthie and under the palaver tree. The head of each district and the Imam settle the contests, plan events to promote solidarity and prepare ritual festivities in such a way that the poorest persons might obtain a piece of meat owing to the participation of all. This is the location of Friday prayers. One or two open spaces can be used for contests and games between neighborhoods or neighboring villages in a spirit of fraternity that never fails.

## Traditional Villages Today

The harmony and social cohesion of traditional villages have been broken down and destroyed from top to bottom. An artificial opposition has been nurtured between the urban Lebous and the rural Lebous. Now the prominent residents of the twelve traditional districts of Dakar are entitled to all the honors, while the residents of the villages remain voters, but are never electable. The traditional authority has been subverted by the appointment of the Grand Seringue by the central administration, over those elected by their traditional constituents. They had to cooperate or resign. Thus the struggle among the different wings of the dominant parties had serious consequences for the traditional societies, which resulted in internal divisions and weakening traditional authority. Neighborhood leaders were replaced by delegates appointed by the administration, who have been chosen according to criteria that reflect goals of the central government. When all the signposts are moving and when the right of appeal becomes undefined, the ability to resist weakens. It is in this context that the National Domain Law, which has ruined the Lebous, was adopted by the delegates, including those from the Lebous.

One of the major resources of the traditional villages is the land, basis of the communal life. How has the land been managed in Ouakam? Located at less than 10 km from the center of Dakar, the village of Ouakam extended over 272 square kilometers, with a population of 37,000 inhabitants in 1995. In 1940, the population of Ouakam was 2,385. In 1960 it was 6,098. At the 1988 census, it reached the impressive figure of 23,413 inhabitants. Based on local birth rates and rates of immigration, it is

Bill Mastin

The road from the market
to the mosque at Ngor.

estimated that it will reach a population of 73,600 in the year 2008. Since the 1970s, this demographic evolution can be explained only in connection with decisions made by the central administration since 1960 when Senegal became independent.

Resting on the republican principle that "the nation is one and indivisible," The National Domain Law (number 64-46, passed on June 17, 1964 ) was adopted. In principle this law is unassailable. In application its goal has been to dispossess the Lebous' land on Cap Vert. The Lebous' properties were the heritage of their ancestors. The process of matriculation and demarcations, to obtain title to a property, was and is publicized in such a way that the concerned individuals, often illiterate, are not correctly informed in time. Even though the occupants might manage to keep their parcels, these could eventually be transferred to the National Domain, allegedly for lack of modernization. Fields of manioc and millet are of no value for developers who can quickly build modern dwellings, on grounds that have been given generously to them by means of a fictitious lease. The apartments are rented at high prices to expatriates. The Gueye family of Ouakam, for example, lost 25 hectares

(62 acres). In spite of various protests and procedures undertaken, no compensation has been given, no member of the family has received any allowance from the central administration in any form.

The decree number 82-438 of June 24, 1982, approving and implementing the project of improvement of the traditional village of Ouakam, was shelved almost as soon as promulgated. The State was far from being able to provide the money required to start the project. The planning did not take into account the fact that the lots that were supposed to be provided for relocation did not exist. The only purpose of the plan was for its use as a propaganda weapon for one political faction against another. While the Ouakamois were waiting for the allocation of lots, the administration gave the land away to the members of a builders' cooperative.

The worst about it all is that those landlords, who claim to abide by a dubious legality against the legitimate common law owners of the land, are often speculative developers as has been attested by what happened to the National Assembly Coop when the same lots have been resold to two or three different persons. To resist these aggressions, to preserve the heritage of their forefathers, to answer an urgent need of owning a decent lodging, and in order to preserve painfully accumulated savings, several families claimed lots for themselves and built homes without authorization or professional guidance. They have been called "spontaneous habitat."

The decree No. 78-124, of February 7, 1978, expropriated an area of about 350 hectares from the Lebous of Ngor, Dakar, Yoff, Ouakam, without compensation and without qualms. In Ouakam, subdivisions were built on fields that used to be cultivated by the villagers. The extension of the runway of the Dakar-Yoff airport destroyed all agriculture and cattle-raising in Ouakam. Truck farming is tolerated and continues in the airport area using waste water loaded with polluted, pathogenic agents.

Fishing does not feed fishermen anymore in waters where prized fish were once plentiful. Frail fishing boats are hardly a match for the modern trawlers outfitted to plunder unprotected fishing grounds. As if these aggressions were not enough, outsiders intrude using explosives, jeopardizing the health of the population and the replenishing of the fishing grounds.

How can we resist these aggressions, manage remaining lands in the most efficient way, use the soils rationally, improve the environment and create new resources? The population will have to answer these crucial questions quickly and correctly because the traditional villages are at a crossroad and, for most of them, their fate is tied to the fate of the big cities.

## Traditional Villages at the Crossroads

The unwise development of the capital city of Senegal poses serious problems of sanitation, hygiene and insecurity that will be resolved only by giving a soul to Dakar. That presupposes that the people tied to this city, who are often related to one another, get together. Their attitude and behavior cannot resemble, in any way, those of the speculators who care more for making an easy buck and for enjoying vulgar pleasures than increasing the prestige of the city as an African metropolis. The

Richard Register

Every year more buildings fall into the Atlantic at Yoff as the waves etch away at the sand. Locals suspect that new development that takes sand for constructing buildings is responsible, or perhaps rising sea levels.

native population, the Lebous, ought to unite more tightly to experience their strength around a prestigious community leader having enough influence and lucidity to sanitize the mores and the commercial circles without xenophobia.

It would be indecent and irresponsible to impose upon the honorable septuagenarians who compose the membership of the Council of the Elders, an agenda as loaded as that of the Government of the Republic. Thus, the Council of the Elders must delegate some of its powers to mass organizations such as the Association for the Economic, Cultural and Social Promotion of Yoff, APECSY, the organization headed by our host at this conference, Serigne Mbaye Diene. The Council for the Integrated Development of Ouakam (CODIV) is another such group. These organization that taught the population to resist land speculators and militate for a rational, equitable management of the remaining lots must become centers of reflection, orientation, understanding and mobilization for all concerned citizens.

The CODIV under direction of Urban Development and Direction of Urbanism and Architecture initiated a more realistic sectorial approach to the subdivision of one of the oldest neighborhoods of Ouakam. That example of restructuring deserves to be followed and supported as a method for involving the population in a process of development that concerns them. The city of Dakar has approved a budget to this effect and the government of the Republic of Senegal, by creating a fund for the land restructuring and rationalization, has manifested an interest in making such undertakings financially viable. The decree 93-522, of April 27, 1993, followed the same approach which led to the elaboration of a blueprint for future urban development. The public and civic centers, as well as the residential areas, have been identified. The residential coop, created by the CODIV is very popular because it responds to an urgent, clearly formulated need.

The problems of implementing the planned development remain. Types of habitat need to be defined, and selection of building materials most likely to provide satisfactory lodgings for the largest number of young people with low income needs to be determined. The goal is to keep these people in their native area, so that they sink roots in the most productive traditional values and participate, in an atmosphere of solidarity, in the struggle against violence and all the other types of nuisances. Today's problems of security occur in our villages with an intensity resulting from mass mixing of populations creating an atmosphere of anonymity that produces a liberalism that often becomes libertine, not to speak of the deleterious influences resulting from the city, seat of the depravation of the mores and damnation. It is interesting to note that there is no liquor store in a traditional village where everybody knows everybody else. Elsewhere, after midnight, groups of prostitutes from Ghana concentrate in the suburban neighborhoods such as Bira-Ouakam, or Airplane-City.

Regarding waste disposal, in conjunction with the delegates of the neighborhood, with the approval of the Council of the Elders and the support of the Constabulary, the CODIV will strive to sanitize this district by all available means. This cleaning up requires better supervision of the Common Land Property of the village on which trucks coming from the city clandestinely unload all kinds of refuse without minding the consequences of these actions on public health. We cannot tolerate that outsiders defile our environment with impunity. Some irresponsible or lazy people find it easier to empty their tons of garbage on the grounds of the village. They will persist in doing so as long as the village will not take drastic measures to preserve the health of the population and as long as it will not warn trespassers of the stiff penalties they might incur.

As the CODIV noted in correspondence addressed to the Director of the Sanitation and Drainage Department on May 3, 1995, the master plan of Dakar does not take into consideration the traditional villages, because they are neither rural nor urban. These villages are purely and simply ignored. For example, the village of Ouakam has no sewer network to get rid of the waste waters, therefore the village ladies have to dispose of it in the streets. To remedy this situation, the CODIV requested the Belgian Cooperative to extend its project of treatment of waste waters to be recycled for use in Ouakam.

In extending this plan the Belgian Cooperative would create locations where the waste waters and the human waste could be treated and eventually transformed into compost. Thus truck farmers would be in a position to use a larger quantity of treated waste water and compost, as a substitute for much more expensive chemical fertilizer. The management of the emptying locations would create jobs for young people. The technique of "ranked mosaic artificial ecosystems," created by the team of Professor Radeaux of the university of Luxembourg, would provide a natural purification which would make possible planting new trees in the area under consideration which, in turn, would provide firewood and stimulate the development of local crafts.

This well-conceived project would have a considerable impact on the improvement of the living conditions of the population and, therefore, on its health. Unfortunately the project has been delayed by the veto of the

Susan Felter

Solidarity has been a high value and effective strategy for hundreds of years for the Lebou. With solidarity they were able to resist slavery and maintain a high level of independence throughout the colonial period. Here, Yoff resident Mamadou Mar shows a mural near the center of his village depicting the slave house on Gorée Island barely a dozen miles away.

government agency whose authorization is necessary to build a water purification center in the Airport Zone. This veto is hard to justify since the truck farmers have been active for 40 years in this area, without jeopardizing the security and safety of the Dakar-Yoff airport.

Let us be united so that we can secure the signing of this document which benefits us all. In a region of the Sahel, where the capital city and its suburbs produce waste waters, of which only one third is treated, the scientific significance of this project ought to incite each of the participants to contribute to its implementation.

To conclude, I would like to say that this presentation and analysis of concrete situations in Ouakam and other villages should allow us to perceive the ills resulting from underdevelopment and, even more, the ills resulting from bad planning. Implicitly we are faced by a societal disease resulting from uncontrolled development of cities and from poor resources management. My belief is that, if the cooperation of the government and the local authorities were achieved, all the problems I have mentioned in this presentation would find satisfactory solutions.

Water tank construction site in rural Senegal.

*"In the last three to five years, there have been new findings in technology and science opening up a possibility for the Africans to invent a [development] system of their own. [They] should be able to develop a new momentum that is independent of the large city," says Richard Meier. Will it be a marriage of traditional village and ecological city?*

# 6. Development –
# Africa and Beyond

In the land of asphalt and freedom, gasoline and smog, we are familiar with various love/hate relationships, as with the car: our friend taking us to work and entertainment, even escape; our antagonist that tricks us into traffic jams or the repair shop for an average cool $5,000 a year. Multiply those countervailing sentiments several times over and picture your whole culture turning into quicksand with deep debts and no economic certainties in sight. Then you can imagine a yet larger ambivalence – about development itself, in all its dimensions. In the previous chapter, the Africans wanted it eagerly, or rather a selective, not well defined slice of it, and found the rest most cruelly destructive of their cherished traditions and values.

In this chapter we look carefully again at development, but from different angles and in different parts of Africa, as described by three organizer/teachers, Gregory Vlahakis, Gladys Elsie Mpatlanyane and Sue Hart, of EcoLink, the inspirational NGO from South Africa . There, the educators don't "educate," but rather "share information" with the people. There, the knowledge of the traditional cultures and growing knowledge of ecology and "appropriate technology" are respectfully brought together so that all participants can choose the kind of development that's best for everybody.

Asks Janis Birkeland, "Has the Western development paradigm brought social security, improved access to food, protected the environment or the weak, eased the population crisis, or improved the plight of women?" She finds the globalization of the "free market" more than a little problematic but traces the outlines of a new development paradigm congruent with aspirations for ecocities and ecovillages.

Zenon Stepniowski from Poland travels to the Comoros Islands in the Indian Ocean offering principles and practices gleaned from around the world for providing low-cost housing – partially by reminding people what they already know and encouraging them to seek practical solutions with local materials.

Does Richard Meier describe a new face to the same old Western development model, or do new information technologies hold great potential for serving people in tribal and developing cultures? After all, though Western development has meant exploitation and destruction of cultures and natural environments, it has also meant vaccinations and food growing and preserving technologies that have saved millions of lives. Meier's cheerful glance into a quickly improving African future, delivered with the assistance of computers hooked to televisions, to satellites, to digital phones, to video game-like simulators for training villagers to fight agricultural battles against plant microbes and insect horeds actually does make sense if.... You decide.

Finally, Joseph Smyth, developer from the land of sprawl, suggests early steps toward building a new ecocity economy. He has a proposal, possibly a way to help transform development world wide so that – at last – we begin to build *the right thing*.

# Our Common Future – or Neo-Colonialism?

## JANIS BIRKELAND

Assalaam malekum. (*To which the audience replied, "malekum salaam," the traditional greeting and response in Yoff and throughout the world of Islam. It means "peace be with you" and "to you peace," the always proper and friendly greeting in Arabic.*)

Since I became an academic I've had very few opportunities for meaningful experiences so I would like to take the opportunity to thank everyone here in Yoff.

Today I would like to question what is called the dominant Western paradigm or mainstream model of development, and contrast that with a new paradigm or movement that is emerging in the developing nations, one of community-building and local empowerment. The basic question I wish to raise is: "does the rhetoric of a 'Common Future' that we hear from world leaders today herald a change toward a sustainable global community or is this the old colonization process in new clothing?" And, where do we as environmental planners, designers and citizens fit into this global transformation?

Now my questions suggest that the conventional Western approach to policy making, to which most professionals, academics and government officials are still wedded, has failed. The development "solution" has become a crisis of monumental proportions itself. I will suggest that, if we do not deconstruct the intellectual and institutional roots of this approach to decision making and development, we may become part of the problem ourselves.

The developing world, or South, includes two thirds of the world's population, and yet is seen by many in the North as being at the margins, as if they were "out there," not connected to the Rich World we inhabit. Many of us view the South as if it were a theater, removed from our own lives, so we can make believe that the horrific scenes we see on TV nightly are not real.

*From a talk and paper by Janis Birkeland, Co-Director of the Center for Environmental Philosophy, Planning and Design, University of Canberra, Australia.*

This is a development scenario wherein Man is portrayed as fighting against the forces of Nature to achieve liberation from natural constraints. This drama, this fantasy has lost all meaning in a world where people now struggle to protect their very means of survival – nature and community – from the forces of unnatural development. We all too easily forget that many have actually been murdered, in "real life," for trying to protect their rivers, plains and forest homes.

Now one might be forgiven for thinking that there is no real alternative scenario to this assimilation of the developing world within the iron cage of industrialism and modernity. Decades of witnessing endless depictions of the hopelessness and despair in the South has created a sense of the immutable and inevitable. We in the Rich World have therefore tried to blame human nature or endemic forces in the South for the failures of colonization. Denial is, after all, a normal response.

But I am becoming much more optimistic now because I do see winds of change, and as the recent Women's Conference in China demonstrated, there is a dramatic "continental shift" in the discourse of development now. The center stage is moving to what has been the periphery. Here in the South, the dominant paradigm of "development" (read colonization) is starting to be supplanted by an old – yet new – paradigm of "local participation and self-sufficiency."

But first, let's be clear about what the development paradigm in the old sense has achieved by asking the following questions of it. Has the Western development paradigm brought social security, improved access to food, protected the environment, or the weak, eased the population crisis, or improved the plight of women?

Well first, by luring people into a global "free market," we have done anything but create social security. Once in the market, these people become reliant for their basic needs upon powerful interests, the very ones who have dispossessed them of their raw materials and food sources. When the village men leave for wage jobs, the

Susan Felter

Woman doing chores in a rural village of Senegal. "When the village men leave for wage jobs, the women must still feed the children...."

women must still feed the children, in some places with nothing to sell but their own bodies.

Has the development paradigm improved access to food? Productive local forests have been replaced with plantations that have destroyed the complex and diverse sources of food and materials for village people. The best produce is often exported and many villagers cannot afford imported foods. This monocultural approach has also led to ecological disasters such as mud slides, the destruction of fish stocks and forest foods, vulnerability to fire and storms, and the loss of sacred ancestral ground.

Has the development paradigm protected the environment? Our technical fixes have caused the "chemicalization" of not only the environment but also humans. Hundreds of millions of metric tons of hazardous wastes are disposed of each year, while the sources of good health – clean water, natural foods and nutrition – are being irretrievably lost. Just as the integrity of ecological systems has been eroded, people's defenses against diseases and toxins have been weakened from this conviction that we are separate from nature.

Has the development paradigm protected the weak? Development has often created victims and then cast them as villains. Laws have actually been enacted which literally define forest dwellers as "trespassers" or "illegal forest encroachers" on their own lands. Squeezed by pressure for land by industrial plantation companies and land speculators, villagers are also often portrayed as violating the rights of those invading corporations.

Has the development paradigm eased the population crisis? Western development has allowed a geometric increase in population, yet has shortened many life spans. It has identified the "cause" of the population problem to be none other than the impoverished Southern "home maker," whose children consume, as Marina said in her earlier talk, one fiftieth to one two hundred and fiftieth as much as do children in the Rich World. Efforts which enhance economic and social security have proven successful at reducing population growth, but these solutions are often ignored in favor of expensive and physical contraceptives. Many women in the South suffer from preventable gynecological diseases, problems related to child bearing, nutritional deficiencies, communicable diseases, and infections – yet the resources for research flow into "high-tech" fertility birth control.

Finally, has the development paradigm improved the plight of women? During Marina's talk I leaned to my neighbor, Sue Hart, and said, "she's saying what I wanted to say, so I'll have to change my talk." And she took my arm and said, "No, say it again." So I'd like to say that it is still fashionable to hear in the North that the solution to overpopulation and poverty is to "educate women," when it would seem obvious that the need to re-educate men is also very pressing. When women have no access to safe and effective contraception, and no control over their fertility or sexuality, the choice is not in their hands. They do not have the right to say "no."

So it could be said then that the Western approach to development arguably fails every important test. The solutions we prescribe from the North have been the

Couple in a rural village, Senegal.

source of many problems in the South. Eurocentric and androcentric notions of progress have resulted in alienation and social fragmentation around the world. The Western development paradigm must be understood to have obtained its universal legitimization through patriarchy, a structure built on anti-nature and anti-female foundations. If we are to develop ecologically sustaining societies, we must recognize the essentially unbalanced cosmology of the Western masculinist and "economistic development construct" itself.

But there is an alternative. A new paradigm is taking hold which paradoxically fosters cohesiveness by nurturing social, cultural and biological diversity. The agents of this movement, as Newman says in Vandana Shiva's book CLOSE TO HOME: WOMEN RECONNECT ECOLOGY, HEALTH AND DEVELOPMENT, "do not presume to speak for others nor attempt to create dependency, but rather support and empower these traditional practitioners of appropriate development."

Increasingly, traditional peoples are refusing to play the victim role, some are even rejecting external aid on the grounds that it is designed to "disorganize, disempower and ultimately enslave." Villagers around the world are developing pride in who they are and will be. And these new theaters are diverse, participatory,

community-based and integral with the local ecology. Each and every person is a central actor in their community.

So what has been the political response to these new movements, these new calls for independence – a value that we regard so highly from the North? It is to use the concept of "environment" to universalize the idea of a "Common Future." Could this serve, if we are not careful, to retain control over what people in the South are beginning to do on their own initiative: reclaim the sustainable access of their spiritual lives and their indigenous cultures? We must take care that our large scale policy initiatives do not serve to entrench outdated goals of modernity, goals to be realized through a Common Market Place. To quote from D'Sousza in Wendy Harcourt's book, FEMINIST PERSPECTIVES ON SUSTAINABLE DEVELOPMENT, "In the vision for a common future, there can be no place for a multiplicity of futures; no place for differences; no place for a plurality of cultures."

Now I began by questioning the Western development paradigm, and I want to conclude briefly by posing two questions for us delegates here today. First, will our energies be diverted to further entrench the global market and power structures, or will we be effective in supporting diverse self-organized and decentralized local community-building efforts?

Second, does the survival of traditional ways of life and their environments depend on the protection of power-based patriarchal relationships, or does their survival depend instead upon an affirmation of the "feminine principle" and a re-balancing of our cultures and our relationships with nature and one another?

# EcoLink Programs in South Africa

## GREGORY VLAHAKIS, GLADYS ELSIE MPATLANYANE AND SUE HART

Karil Daniels

Gregory Vlahakis

### Gregory Vlahakis

When we talk of ecovillages there seems to be a general consensus that this is a new concept of living. But in Africa and other areas of the world this has been a way of life for centuries. *(Loud clapping rolls through the audience.)* Recent history has seen a forced translocation of the indigenous peoples of Africa and of other parts of the world that has altered this ecovillage lifestyle. Traditionally, as a matter of survival, the peoples of Africa adapted to some of the most inhospitable environments imaginable.

Our foundation, EcoLink Environmental Education and Development Center, is involved with the plight of the rural peoples through our programs such as Sustainable Earthcare and Food Systems. We work on social policy and water sanitation, extending and improving existing rural technologies, thereby honoring the needs of the people, and not imposing ideas, which in most cases, are alien to rural dwellers.

The role of the community is of paramount importance for this program to be successful. It is important that they participate in all stages of the program, including the design or adaptation of new or existing ideas and the extension of the project over time or into new activities.

The community chooses two members from the village to participate in this program. They in turn will each select five capable colleagues to assist them, therefore contributing to the multiplying effect. The training includes the development of environmental awareness and management skills.

---

*Gregory Vlahakis, Gladys Elsie Mpatlanyane and Sue Hart all spoke for their organization, EcoLink, and so they are included here together. Vlahakis is Director of EcoLink's Sustainable Earthcare Program. Mpatlanyane, also with the Earthcare Program, teaches basic cooking, trench gardening, recycling and environmental education. Sue Hart is founder of EcoLink. All the following were transcribed and edited from talks recorded on video tape.*

The community is encouraged to form committees for undertakings that will require considerable and careful thought at the planning and decision stage. They are encouraged to hold regular meetings to check on progress in the training of their colleagues and the progress of the project at hand. People who show the most commitment receive generous support and are encouraged to upgrade themselves. Involvement of influential people from different organizations is also encouraged. For the integration of traditional African wisdom into a future ecological rebuilding program, we also take into account the different structures, cultural background and beliefs of the participating communities.

It is also essential that women play an equal and in some cases a much larger role in the decision-making process, since they have had to work within a framework where decisions have been made without their involvement. We begin by investigating and testing simple techniques for upgrading existing technologies. This is a first step toward building involvement based on existing traditional practice.

The training process for program participants working in our projects benefits communities by building or acquiring facilities and equipment and by restoring natural resources. For example, there is a very serious problem of erosion in Africa. Land has not been maintained due to lack of community participation or training. Therefore it is essential that we address this problem of erosion by training people to protect and build soil. Thus we are growing grasses which we use as erosion prevention. Simple methods can lead to a lot of improvement of such technologies which then have a wide application through our countries.

We, too, must realize that what may be appropriate elsewhere may still be complicated in our environment. In the past it has been taken for granted that when we get something from, say the United States, which may be appropriate there, that it would be appropriate in our communities as well. But we are still living in very hard times. And it is only now that changes are taking place, especially in South Africa. This is why we have to be very careful how we pass this kind of information to people. In the past everything was brought in and people were asked to work in a foreign structure. It was unfortunate because no one was able to understand how or even whether it was beneficial to use the technology, knowledge or advice.

The traditional belief in the spirit of the Earth has made it easier for us to interact with rural communities and exchange ideas about our one Earth, the only Earth we know.

With a new freedom in our country we will become one people with a new understanding. We cannot live without each other. Now, what we need is "to become one people in the one world we know. In the spirit of the Earth we will be whole."

### Gladys Elsie Mpatlanyane

Throughout Africa, our generation today suffers from diseases that our grandparents never knew. The emergence of diseases like diabetes and high blood pressure is the result of the so-called improvements to our natural food systems. This was done through the addition of colorants, preservatives and flavor enhancements – ingredients like MSG. Food poisoning sometimes occurs when canned and bottled foods or drinks expire, but are accidentally sold to people who unknowingly buy them. This growing dependence on packaged food leads to hospital bills and results in a sickened society that depends upon medication. What makes the situation worse is that the prepared food is really expensive, and is regarded as luxury. Consequently, for our modern, socially developed society, healthy food is a thing that only a few can afford. The solution to the problem is for our people to remember their roots and go back to their cheap, yet nutritious diet.

Karil Daniels

Gladys Elsie Mpatlanyane

Without any formal education, our people knew how to cook a variety of well-balanced and healthy dishes. Unfortunately, people seem to think there is something wrong with their traditional foods. Our duty as community helpers is not to dictate, but to remind the people of the wealth that is right in their back yard. Even with the little land they have and the little water that is available, people can establish sustainable food sources and have enough left to make a little money.

There needs to be a return to more traditional ways and avoidance of foodstuffs which are unhealthy and disease-causing. We work to help make people aware of their cultural roots and traditional ways. We encourage the exchange of ideas on the storage of foodstuffs, using traditional methods, such as the drying of vegetables. We give training on nutrition and organic vegetable growing using the trench gardening method which uses recycled materials, making a sustainable method of gardening. We also encourage social and cultural development. We need to move away from developing program curriculum in isolation and need, rather, to be designing it around traditional systems. It must present knowledge and beliefs which are more applicable to the lives of the people today.

Though not described in detail here, we also include AIDS awareness as an important component in the health program and its target is youth, especially women. Our aim is to enable and empower rural communities to take command of their lives and resources, especially the growing of nutritious food; to assist and improve the community's well-being as they define it; and to encourage the expression and sharing of what rural people already know.

### Sue Hart

EcoLink's work in South Africa, like similar work around the world in recent times, grew from a long and painful process of recovery, or rediscovery, of finding the freedom to make bonds and to have choice in one's life. And so our program has many aspects that nurture the human spirit so that we may find hope and self-respect in the act of building a new nation and a new kind of society of equal justice.

Our own team is a mini-community and we have this dream of an ecovillage. We are fortunate to have a great deal of land and we look at the Swaziland Mountains.

Our team of fifty-five is an extraordinary, unified, committed team. It is a privilege to work with our people. They reflect the courage and the commitment and most importantly the hunger for learning. After starvation for so very long, people hunger for the opportunity, which is after all a human right, to have information and learning.

It has been said by a famous South African that the rural people and especially the rural women are the spine, the backbone of South Africa, and though we understand and accept completely that there is an enormous traditional wisdom, an Earth wisdom as we know it, these people come to us and want to catch up and understand what is happening, what is the state of the world, what is the state of the environment that affects their everyday lives. And we feel that upon the ability to supply the information, the skills, the materials, the assistance, the human relationship, rests the future – the ecological, sustainable and sound future.

We ask many questions and we have workshops in order to try to solve them. Everything we do we call "interactive." Even the word education has grown to have a bad meaning in South Africa because it has always been an imposition of information and knowledge, not an interaction or not a drawing out of what is already there which is after all what should be the meaning of true education, especially environmental education.

We produce in response to this, in several languages, different sorts of resources and projects. Then we evaluate these and assess them with the targeted people and we find we often make mistakes. Then we have to reshape and rewrite these resources and projects.

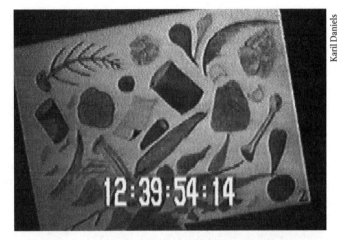

Video of hand-held "Gifts of the Earth" posters. Companion pamphlets from EcoLink are printed in both Zulu and English.

EcoLink is entirely dependent upon sponsorship. We have in the past always been know as an NGO. But since the miracle of our new government has changed our world, the government's own Reconstruction and Development Program has declared EcoLink a program which supports and has effectively carried out programs based on working with the people, always with the people, listening to them, working with them.... We need to have their participation. This is incredibly important. And if one doesn't do this one cannot possibly ever succeed.

We call our program "Growth for Life." It has two separate spokes or aspects. One is personal growth, growth for life on the personal level and the other is the shared community growth and learning.

Our program is for schools and communities, for leaders and the public sectors. We even train heads of corporations and industry because those people at the top don't understand the principles of living and the trouble spots of our environment today.

Our vision for the ecologically sustainable future which this conference is about, and for the creation of the dream of our own ecovillage – we have already started the design there – must depend on the broadening of our horizon. Here are some factors that summarize:

Susan Felter

In the streets of Yoff, on almost every block there are murals depicting local and religious heroes, scenes of village life, historic places and events, health information, political slogans and other items – all deftly rendered by hand and brush.

The vision depends on attitudinal change, on environmental and ecological understanding and knowledge. Please, I want to differentiate between the two terms. Understanding is an epic, emotional, intuitive thing. It is a beginning, and then knowledge and the thirst or hunger to seek knowledge, to research follows. It also must rest on self-help and hands-on development and action. And finally success depends upon continuity and sharing. You can't just help people, work with people and drop them and go away. You have to be there for them, to guide them, to work with them. Otherwise you have frustration, failure and often you have anger.

In order to optimize this program, we have put out some resources such as posters and pamphlets which we use in our interactive workshops. In the country we use hand held posters that we have painted in bright colors, often with food plants pictured, people engaged in agriculture, or natural animals and various cycles in nature, all drawn by hand.

Though we are rural, we move into the city with some of these resources. We also have a video, overheads and so on. We all know so well how exciting a city is and can be, and you take a child or a teacher into training and look at one little spot of the city and what you discover is amazing, and changes through the seasons. Even though some of these educational materials were made before the change in South Africa, we noticed even then how enormously important, how inspiring to the young people it was to clean up rivers, to discover life, and to go into partnership with nature.

Our series of posters and pamphlets is in two languages, in Zulu and English. It is called "Gifts of the Earth." Natural resources and biodiversity are the subject matter. The lessons that emerge are told in the form of stories that we look at quite practically. We ask "what can we actually do about it in our personal lives? How can we take responsibility for our local environment, for our region, our nation and most of all, now that the curtain is up, the barriers are gone, how can we become members of the global community?"

One of our very earliest teaching aids we called "Health of Life." It became very important and focused on the owl. In our country the owl is an omen of death or ill health, ill luck. Where I come from in England the old owl is a symbol of wisdom. In South Africa today it is believed that there are 100 thousand owls left, and if you killed them all you would have certain results. More than half of the maize crop upon which millions of people subsist and survive would be destroyed.

Now, we travel around the world to see how the owls are, in our workshops. In the process we have the opportunity to teach about geography. We can discuss health and see how rodents can bring disease and owls can control the rodents. We have mathematics. We can take this approach from primary school into high school. But when we come to conclusions, in our workshops, we say to our teachers, our students, "all right, this is EcoLink's idea. Now you can use your own resources and imagination for your own condition and circumstance." And we have wonderful results. So our workshops are interactive. We say, "this is something we have done, what would you like to do and take back to your school, community or family?"

We do have procedures and strategies of management in order to produce these resources. We have planning, very careful planning. What do you need? Who is it for? What space, what money, what topics? What is the after care or the continuum? And so on.

I have a particular love for a time which was before EcoLink, and a very bad time in South Africa, in the early 1980s when I had just came back from East Africa and was asked by the first teacher training college in our region near the Kruger National Park to try to create a new way of teaching environmental education. I had 90 volunteers – student teachers, politicians, volunteer teachers, all sorts of people. We discovered together there was only one way to release and realize the human creative potential. It's as if we live with a stranger. We hardly know who we are. And I challenged my students, who were having a very bad time. They were angry, they were frustrated. There was resistance, there was violence. But these people, men and women, were allowing me to work with them to discover that we are all part of one planet. We can look to the future. I'm talking about the time between 1981 and 1985; you can imagine how hopeless everything seemed at the time.

And so in their second language – their first language was Swahili or Zulu – the students were challenged to write about life, not necessarily nature, but about life. And they responded most beautifully. Their grades in college went up, their lives changed because they discovered this potential.

I would like to end by reading you a very short writing by a man who is a school principal in South Africa. His name is Lucas Niecy. This comes from a book called CRIES OF THE HEART which was finally published. This is a very special poem and I have never stopped reading it. It is called "Night Coming Tenderly."

At such a time black Lucas can look at the
    star-lit skies and find that he has, after all,
    a place in the universal order of things.
The stars, the black skies affirm his humanity, his
    validity as a human being.
He knows that his belly, his lungs, his tired legs,
    his appetite, his prayers and his mind are
    cherished in some proud involvement with
    nature and God.
The night is his consolation.
It does not despise him.

# Urbanization Everywhere and Low-Income Housing on the Comoros Islands

## ZENON STEPNIOWSKI

To understand the current situation of human settlements in developing countries, it is necessary to know the socio-economic and demographic situation of the world. We need to take into account the accelerated growth of human population and the universal urbanization that is resulting – one of the most important factors in transformation of housing in the history of humanity.

Analyzing the growth of population reveals the scale of the problem. Some summary numbers: Around the year 2000, for the first time in history, urban population will surpass that of small towns. Of 6.9 billion people on the Earth then, 3.6 billion – well over half – will live in cities. In the year 2025, the number of inhabitants in cities will be equal to the number of the population in the whole world today.

The growth of the larger cities is particularly noteworthy. In 1960 there were 114 cities in the whole world with more than a million inhabitants. By 1980 there were 222 cities of over one million. Today there are approximately 400. If the current trends continue, by the beginning of the twenty-first century, 22 cities will surpass 10 million inhabitants each. The bigger ones – Mexico City, Sao Paulo and a few others – will approach or surpass 25 million each. We can estimate that these agglomerations will be the poorest cities in the world, having a great number of squatters and homeless.

This growth of urban and total world population provokes disintegration of rural structures due to the growth of the rural population in some areas, and in other areas, the abandonment of villages and rural areas. As rural centers break up, subsistence agriculture becomes difficult or impossible. Lands become more and more arid. Secondary cities and their social services weaken and regional administration and job markets force rural people to move to the larger cities.

*Zenon Stepniowski is an architect and town planner with the Institute of Architecture and Spatial Planning, Polish Academy of Sciences, and consultant to the United Nations Center for Human Settlements.*

In the urban zones, construction is not even foreseen. Technical and social infrastructure and town layout are neglected. As a consequence, poverty and wild, chaotic, inhuman conditions result. The crisis of the city is magnified by the demographic concentration, the number of inhabitants grows without limit, the degradation of the human environment develops, and life in this degraded milieu becomes very difficult, if not impossible. In general, this phenomenon presently afflicts most cities, but above all, those of developing countries. Here and now we need to assure shelter for a billion people.

Traditional urban structures deteriorate in relation to uncontrolled growth; the migrant population grafts onto the existing fabric or on lands near cities. These areas are unhealthy and inadequate for construction. The acuteness of the housing crisis is noticeable in the density of such sectors, accentuating the deficit of social services and equipment. Even though the solution for slums proves to be generally temporary and the quality of life offered is not worthy of a human being, such a way of building shelter does not cease.

The old infrastructure becomes overwhelmed by the influx of population. Many difficulties and total impoverishment ensue. In addition, the lands laid out spontaneously possess little or no infrastructure, and the deterioration of the human environment accelerates the degradation. But there are principles for guaranteeing healthy spontaneous urban growth in developing countries. Though the "occidental" methods of urban planning are not coherent with the problems of developing countries, on-site research has permitted us to define the principles of urban planning in these countries.

1.) The planning model should be subordinated to the principle of "historic continuum," compatible with the traditional and natural environments; the relationship of man to ecosystem must be respected as a factor of primary importance. 2.) Spatial planning should be simplified for the housing but 3.) the construction of social facilities, commercial and cultural centers and of small, productive businesses is necessary in order to

New housing, Mormo-Hadoudja, 1987.

Zenon Stepniowski

create "attraction poles" for the people and economy. 4.) Systems for water, electricity, roads, sewage and garbage disposal need to be established or improved. Meantime, 5.) there must be coordinated simultaneous socio-economic development of rural areas to deter immigration of farmers to the cities. And, 6.) lots have to be legally assigned to people in areas "invaded" by squatters.

I don't pretend that these principles for balanced development of human settlements are exhaustive. My goal is simply to reveal the problems that will decide the future of our planet.

Low-income housing on the Comoros Islands exemplifies an approach to housing that uses the above principles. For general orientation, Comoros Island is formed of four islands grouped halfway between Africa and Madagascar. Their total land area, approximately 1,000 square miles, is about the same size as the Azores and about one sixth the size of Hawaii. The birth rate is relatively high at 3.6% increase per year today; from 1985 to 1992 the population is estimated to have risen from 395,000 to 600,000.

The government of the Republic, considering the need for housing to be critical, has made an appeal to the United Nations Center for Human Settlements to assist in resolving this complex problem. The UNCHS strategy in such circumstances is to apply the necessary means for improving living conditions while building housing, plan for urban growth while stopping degradation of the human and natural environment, put into effect appropriate legislative norms, and educate professional national managers.

The limits of this presentation do not permit enough time or written detail to explain work as complex as a major social housing project. It is appropriate, nevertheless, to note certain observations. We first studied the general problems pertaining to traditional housing in Comoros, looked at demographic evolution and the socio-economic structure. Then we studied climate, physical conditions and the technical infrastructure, looking carefully at the ecological impacts of building materials and processes, taking care not to destroy natural resources such as wood, coral reefs, beach sand and so on. The management plans of the principal settlements have been resolved, thus defining the guidelines of a strategy for long term urban development. At the same time, on the basis of previous studies, some experimental housing has been built. This method, has permitted the definition of all the elements according to the system of evolving housing, designed for low income population.

The financial structure foresees the participation of the household by way of a modes initial payment. The very low monthly payments will be paid during a period of ten years. The candidate for the social housing receives a parcel of 150 square meters surrounded by foundations (four modules, each of 9 square meters) already built by the developer and this with an outhouse and easy access to the technical infrastructure of water, electricity and streets. The candidate can make a choice of the type of housing relative to his or her income: either the candidate constructs his or her own housing over the existing foundation or the developer provides one, two, three or four ready-to-occupy rooms according to the individual's needs .

# Hopeful Patterns of Development for Africa

## RICHARD MEIER

Richard Meier

As I speak, internet computers are being installed in Yoff. They represent a technology that is hardly known to Africans, and the traditional people of this village are being introduced by APECSY today to the internet at the APECSY office. But this is only the beginning.

Now I will try to describe how that information revolution may speed up change to create a new kind of development in Africa.

I'll start by saying that the European approaches to economic development, including American, were taken to Asia after World War Two and they did not work. For ten or fifteen or twenty years in different countries, nothing seemed to work. I went to Asia to find out what was difficult. I found that they had to invent their own method of development, and usually the Westerners – the Western economists – did not know the local Asian language, so the Asians invented something else. They invented several different Asian approaches to improving the GNP of their own countries. As a result, within an extra ten or fifteen years after they got going, they now have the record growth rates and change rates in the world. They are changing sometimes in the wrong direction, but they have the momentum so that they can change away from the automobile, for instance, when the price of gasoline goes high.

So the Asian societies have invented their own systems and it is not very well known in the economic text books or anywhere else. I had to ask the Asians "what did you really do instead of what you told your advisors?" And this was an eye opener.

What they did, really, was make a city, a city like Dakar, an engine for development. The city then pulled development after it as if it were some kind of a train. And some provinces were slower than others, were

behind others, but the cities generated the development, the large cities, the metropolises.

When I had my own students look at what could be done in Africa to use the Asian experience we found that it would not work. Neither the Western idea nor the Asian idea of development would seem to work in Africa.

Now one of the things that happened in Asia is that the population growth was quickly reduced as soon as literacy became common in the population.

In Africa the extension of literacy does not seem to reduce the population growth. Therefore something more than literacy is required for Africa. The Asian countries were able to increase the GNP growth (after they learned how) by 5% to 13% per year. China has been growing at 10% to 13% per year for the last few years. So they had a kind of formula for moving the society starting from the metropolises.

The Asian approaches were actually tried in places like Egypt, Ghana, Kenya and Nigeria for a while, and they did not work. They didn't seem to fit any kind of African interests or capabilities.

One of the reasons is that the population was growing too fast. As soon as you developed a surplus that should have been invested in infrastructure it was needed to feed the children. Another feature in Africa is that it had less education to start out with. Any attempt to increase this education with international help was overwhelmed by the number of children that came along requiring education. So this population was a serious thing.

But other things, water, drought – these were very serious too. And stress induced by the other factors induced political disorder – yet another issue.

The World Bank didn't know what to do. It felt that some kind of discipline was necessary in the handling of money. So they asked the International Monetary Fund to introduce some governmental discipline in order to

*Richard Meier is Professor Emeritus of the University of California at Berkeley in the Department of City and Regional Planning. He has studied Africa and traveled the continent extensively over the last thirty years.*

gain some kind of credit. The result was that there seemed to be no hope for African development among the international agencies.

But now, in the last three to five years, there have been new findings in technology and science opening up a possibility for the Africans to invent a system of their own. It should be able to grow very rapidly. It should be able to overcome the population problem, it should be able to develop a new momentum that is independent of the large city, even though the large city will continue to grow. It does not depend upon the large city as the engine. The engine can be in the whole society. And it comes actually from above, from the communications satellite. Information can flow into and out of the village quite far from the city.

So what I am suggesting is a way of using information technology to solve the food, the water, the education, the family planning, the organizational, and the political stability problems, and a few others on the side, by using information as the tool.

The new findings I should at least name so you can recognize them. The communications satellites will be overhead in two to four years. There will be two sets at least, one coming out of America and another coming out of Hong Kong. They will be competing with each other and will have a tremendous excess capacity to handle information.

Another thing that is happening is barely known in America right now. It's called the digital telephone, or the personal communicator. It is a portable telephone similar to the cellular phone but of higher performance. It is connected to the communications satellite or to some kind of station on a mountain top or a tall building in a city.

Another thing that has been learned is that we can now transmit the kind of information needed in education and education can be accelerated. What previously took five days to learn now takes, perhaps, two days. Now the children can't digest that fast. So what that means is that the other three days the children can be

A construction site in Yoff.

learning other things in school subjects. In other words you can use this technology for speeding up the essentials of education and then they can be learning other things in addition.

The family planning discovery came only a couple years ago when two different kinds of studies came out, both of them, it turned out, in different places in Nigeria. Then they were confirmed in six or seven other places in Africa. And that is that whereas in Asia family planning could be introduced quite effectively with two to four years of education, particularly of women, in Africa it takes something like six to ten – a middle school education. It needs more and we find this is true of tribal societies in Asia also. There are several hundred million people in tribal societies in Asia and they also require more investment in education in order to bring about the voluntary acceptance of family planning.

The other feature of this is that Americans and Europeans always focus on the women and the women said "you ought to talk to the men." And so finally a study was started in 1990 and reported last year. They produced a radio program based on the life of a man in the village or market town, and the life of the man was made more difficult by the demands of extra children. And therefore they had a series of problems that the men faced. Quickly they began to see that problems were always complicated by too many children. Then four years after they started the program, they started asking how many children were the men insisting upon in the family and it had been reduced from five to six down to about three and a half. So there was a very consistent reduction through mass media because of the judgment of the fathers as well as the mothers as to what was the appropriate size of the family in contemporary Africa.

Another has to do with – it is called – "expert systems." How can we get the information we need to the farmer, to the producer *just in time*. The Japanese have been saying that their manufacturing industry was based upon this just in time principle. It appears now, with the communications satellite and with the digital telephone and a few other things that we can deliver the information just in time, not just to the farmer, but to the person who manages the water, to the person who provides the fertilizer, to the person who provides other kinds of supplies. So that it all can be coordinated through an expert system that guarantees that people are told about the needs just in time so that they can all be fitted together to increase the size of the crop.

One other feature we ran across was that books in Africa decayed very rapidly, that they fell apart, that they did not last, and that it was hard to replace them. But now there is a new development that is much cheaper than the books. Where the book cost $5 to $20 a piece, the CD ROM disk costs less than $1 and – and it contains the equivalent of several books on it. You can put a whole encyclopedia on a single CD ROM disk. Therefore we have the potentiality of bringing much more information to the learner, to the student through this particular device. There will probably be an improvement in the CD ROM five years, eight years from now, but I'm talking about real things that exist now and are expanding rapidly.

Another is the simulations that can be put into a computerized form. For instance in America and in Japan, if you want to learn how to drive an automobile they don't trust you with a truck or an automobile, they put you on a simulator and a television screen is connected to the controls and you go through these synthetic environments and you learn how to drive before you get into the vehicle.

Now there can be many applications of this but I'm particularly interested in the simulation of growing crops. How can we teach people how to grow crops before they use up the water or the seeds? So we have a new development that will allow simulations and simulations can be put on the CD ROM disk as well.

Let me say how I imagine this might be. There might be a truck that goes out in a rural area, and on the back of the truck you have a series of chairs facing a television set which is connected to controls something like a video game. But these people are going to be growing a new crop – in my paper I describe growing corn. In Africa the corn is grown 70% to 80% by women. So we think of this as a language that is suited to the women and they can work it out, even if they have not yet had a very elaborate technical education. If they have literacy and if they can understand pictures they should be able to deal with the simulation for growing corn. I grew up in corn country so I had great fun trying to describe how one might grow a big crop of corn. In fact I worked my way through college with hybrid corn.

Another that is coming along is a combination of a computer and a television set. When those communications satellites arrive overhead it will be very easy to tap into programs from elsewhere in the world on television

Young men in Yoff. "Any person who feels he belongs to the community, if the community has solidarity, would have a high level of happiness." Richard Meier.

sets. I've seen just a few that do it already where they have antennae on the roof right here in Dakar. But it will be much cheaper in the future.

So these combinations are just now appearing on the market. They will expand very rapidly. You will probably have a factory of your own in three years in Senegal.

All this that I am describing is information technology, and the reason I can make predictions is that I live in California on the edge of Silicon Valley, where about three quarters of all the innovations that are coming up in information technology are centered, so we hear the rumors at the University of California about who is doing what. For instance the last rumor I got just before leaving Berkeley was that the digital phone was going to be cheap to buy. The superior product, much better than the cellular phone, would be available at around $45 by the year 2000 in America and maybe by the year 2003 here in Senegal. It is coming very fast. It is now $500 and will come down to one tenth of that. So at $45 a group of families in a tribal or village situation, will be able to afford a digital telephone.

This is sort of magic. It's really an amplifier of what you want to do, and not all of these things will be done simultaneously. You can select from them and learn about them and make use of them. Here is a potential that never existed before, that Asia didn't have. It is now developing with its own information technology very fast.

Now what are the things that can be done? I've indicated the problems of food, water, education, family planning, building of organizations, and political

stability. I should indicate that these are related in many ways. For instance, we have found looking over the whole world that when political stability is discovered in a country, it follows upon education. What happens when people become educated is that they self-organize in more sophisticated organizations, and those organizations have to bargain with each other and in the bargaining you find ways of getting around the authoritarian approach. And then you begin to have a kind of stable group of leaders in the society of small, middle and large organizations and this reduces the amount of violence in any kind of change in government. This is the observation that has been made all around the world, that once self-organization which comes out of education gets established, then you get a rapid reduction in violence.

What does one do with this information? What are the implications? One of the things I've discovered is that the kind of development that occurred in Asia produced GNP growth and economic security. But people were no more happy, sometimes less happy than before. The level of living was still very poor and, therefore, their method of economic development was quite unsatisfactory. So a group of social scientists went around last year in Asia and tried to find out about happiness. They asked people, "are you very happy, quite happy, so-so, unhappy, or very unhappy." And they came back six months later and asked the same question. Unless there was a death in the family, the loss of a job or something like that, they were really the same as they were before so long as the person's life situation did not change.

Ihlara, Turkey. Similar rural villages around the world are being emptied by economic pressures, their people migrating to the city.

A few miles outside Afyon, Turkey, gigantic housing blocks typical in many parts of the third world are being built. These will soon fill with people dependent on cars and long bus routes.

The question was how can you use a society to produce happiness. We just had the argument about solidarity. It appears that here in Senegal solidarity and happiness are very closely related. And this is something – though the studies have not been done on Senegal – but I would forecast that any person who feels he belongs to the community, if the community has solidarity, would have a high level of happiness.

The quality of life in terms of access to education, relatively comfortable housing and sufficient food – those kinds of things are associated to some extent with happiness. So we have two things: material quality of life and a sense of happiness as being two things you could start thinking about as accomplishing with the aid of the new technology. We found out from the reports that happiness was rather stable, but then we asked "how can we produce more happiness."

We found that it was different in different societies. In America it turns out that if you want to change things, the people are happiest if they participate, even if their view point is not taken. In other words, they make a contribution but if the decision goes the other way, they are still happy that they have participated.

The final point that I have to say is that among the Asian societies, those in the Philippines seem to be the happiest. Yet they are not yet sustainable because they have not found a method of reducing the population in the Philippines. They have quite a bit of education but they need more education and more of the right kind. Therefore the population issue over the long run will determine whether they can stay happy or whether the pressures of famine will catch up with them.

So for ecological planning, and for design by architects, one would have to say that we have to consider sustainability over the long run and this includes numbers, resources, water and everything. I have only given you a hint as to how we can save food and produce water. But all these are possible now with the new information technology.

# Getting Started – A Possible Economic Strategy for Ecocity Development

## JOSEPH SMYTH

Joseph Smyth at Rastenberg Castle, Austria.

If we ask ourselves "what is the nature of our destructive development patterns and how can we make them healthy," we might begin to comprehend that there is one overwhelmingly large impediment and one overarching guide to success. The impediment is the automobile-oriented way of building cities, humanity's largest physical endeavor. The guide to a healthy future is understanding the principles of life systems and valuing the other living things themselves. From these two realizations we can comprehend that we have been building the wrong thing, which in turn largely runs our lives; we can see how to begin structuring a new economy based on ecocities, restoration of natural land and land for food and forest products.

Before people were on the Earth in great numbers, and especially before the agricultural and industrial revolutions, the natural systems of the biosphere were whole and functioning well. The air, water and soil were clean. Life forms were abundant and diverse and all were living in harmony and dynamic balance – a seamless web of life moving to natural rhythms of day and night and the yearly seasons. We can only imagine this original beauty. And underlying it, there was a natural economy in which every actor in the biosphere took something for its own life, lived off the flow of energy from the sun and recycled its own constituents back into the soil, contributing just a little more wealth to the life that followed.

Things on Earth have changed. The air, sea and land are now thick with pollution. Every aspect of the biosphere is being altered, damaged or destroyed at an accelerating pace. Extinctions, overpopulation, and overconsumption are expanding exponentially, and the destruction of life-sustaining systems is a dominant pattern on Earth today.

*Joseph Smyth is an architect and developer from Southern California. Here he presents a few of his ideas from a slide show rich in his own drawings and computer images of transforming the city of Los Angeles.*

At the same time, we know more about the Earth and its systems than ever before, and we know more about ourselves and one another. The race is on. If mankind is to survive, we must make the conscious creative choice to protect, preserve and restore the Earth's life support systems. If the conscious creative choice is made, what will it look like? How do we get there from here? How do we pay for it?

Clearly we must re-vision our life on the planet and in our communities. We need to set forth the following vision, to be expanded and refined. We need to 1.) develop sustainable lifestyles in urban and rural communities, 2.) live within our means and not mortgage away the future, or reduce choices for future generations, 3.) recognize that human beings and human habitat are a part of the ecosystem and its beauty, 4.) accept that cities are as natural to human beings as the biosphere is to this beautiful planet and that they are an integral part of our own creative evolution as a species, and 5.) use democratic, collaborative processes to shape the future of our cities and our world in exciting, harmonious and meaningful ways.

This vision will unfold through the careful, conscious building of an ecocity economy which can come about as – and only as – we go about building the ecocity itself, on all scales from hamlet to big city.

Economic theoreticians in this field, like entrepreneur Paul Hawken and architect William McDonough, see applying recycling to its ultimate degree as a key, perhaps *the* key economic concept for the future – as it appears to have been since the beginning of life's economy on our planet. Richard Register proposed earlier in the conference what he called "four steps to an ecology of economy" as an economic strategy that takes the rigors of recycling as fundamental, then lays out steps starting with ecocity mapping, then listing of the technologies, businesses and jobs to build what's represented

United States NASA

Before: aerial view of "gossamar thin" Los Angeles, California.

on the ecocity maps, then writing the incentives to help it happen and finally recruiting the team of people to build the alternative.

The big block, though, is getting started – and paying for the early projects. In preparation for that, we also need to focus in more closely on the nature of the present city as a problem, see its "vulnerability" to positive change, and understand better the right thing to build, the ecocity. In its most stripped down essence, that is simply the clustered community.

Clustering communities preserves open space in rural areas and can restore open space within existing cities. Clustering also encourages living within a safe, pleasant walk of work, schools, shopping, services, parks, recreation, and public transit. Density can be positive, fostering community and a sense of place.

While the social and environmental benefits of clustering communities are becoming clearer, the economic benefits still remain hidden. The truth is the economic benefits of clustered communities are so massive and so comprehensive they seem *too good to be true*. The tip-off clue in this treasure hunt is that 40% of development's "initial cost" goes to pay for freeways, streets, stop lights, parking lots, driveways, garages, parking structures, and the land they cover. And that's the cheap part. Wait until we get to the "operational cost" – now we're facing the ultimate bottomless pit! The automobile-centered infrastructure is destroying everything dear. It's destroying the natural environment through pollution and urban sprawl, draining our bank accounts, and eating the heart out of our communities, turning them into faceless, high speed traffic patterns and parking lots, forcing us into isolation. If the truth be known, automobile dependency and all of its side effects is the root cause of local, state and national insolvency in my own country and I suspect in most of yours. It's time the true cost of living in automobile dependent communities is known.

Let me emphasize the true dimensions of the automobile roadblock with a thought from Paolo Soleri:

"The re-ignition of cultural evolution will begin as we turn the corner on automobile dependency and reintroduce clustered, pedestrian-oriented living," and a more visceral comment from Le Corbusier: "You go out and the moment you are out the door, with no transition, you are confronting death: the cars are racing by."

Most importantly, it's time for the economic power of sustainable development to be discovered. An early step is realizing how little – in the literal physical sense – is in the way, how little is really there.

That vast metropolis covering literally thousands of square miles where I come from in Southern California is extremely vulnerable to swift change. The enormous damage from earthquakes and fires that sweep the area every few years only hints at the foundation reality of that scattered infrastructure: it is gossamer thin! When we look at photographs of L. A. from the air – on smogless days when we can see it at all! – we need to remember that most of it is asphalt and concrete only three or four inches thick, lawns with roots only a little deeper, and mostly one story buildings. It is radically vulnerable to being transformed into something that could be quite good and vibrantly healthy: a galaxy of compact farm-fringed ecocities and ecovillages scattered like islands across the kind of beautiful landscape of sage and desert sand that existed there just two generations ago.

Thinking through the transitions, and seeing the hazy outlines of a distant new economy based on ecology, I can propose a start in the right direction: we need to build models, models of the sort that Ecovillage at Ithaca and Yoff aspire to be. But we need larger experiments, and many more of all scales and in all different parts of the world. How to pay for them, how to get the ecocity building enterprise going?

While attending the 1992 Earth Summit in Rio de Janeiro, I came across a technology that I believe will prove to be a sustainable funding source for building ecological communities. This technology is a farming process which combines growing timber with food and secondary crops as an understory. Permaculture practi-

Joseph Smyth

After: drawing of Los Angeles as islands of cities, towns and villages in a sea of natural and agricultural landscapes.  Here, creeks are restored, natural areas and wildlife corridors are re-established.  Transit replaces cars, rail lines rise over or dive under wildlife corridors.

tioners like our speaker Thomas Mack describe this system as "agroforestry."  It is well suited for use in semi-arid regions, because the tree canopy creates a protected space underneath – a modified micro-climate – which makes possible the growing of a wide and changing variety of crops beneath the trees as light levels change over several years.

In particular, I am interested in growing the Chinese powlonia tree in an agroforestry system.  The tree is very fast-growing; it grows to the height of a person in about four months, and reaches four to five times that height within a year.  A five year old tree may have an eight inch diameter and is seasonally covered with purple flowers that bees love to turn into honey.  The wood is very useful; it is light weight, but fairly hard.  Though not of construction industry timber strength, it is excellent for many containers, cabinets, furniture, toys, educational materials and musical instruments.

I see an urgent need for jobs and income in the traditional villages around Dakar.  My hope is that you would consider growing trees like the powlonia – though not necessarily the powlonia in particular since one or several others might be better suited to your environment.  The idea is the same in any case.  The reforesting of Cap Vert could then commence – while also producing wood and a wide variety of foods.  I notice that though the streets of Yoff are very quiet and pleasant, there are very few trees and there seems to be little commitment yet to planting them.  Perhaps the cooperative nature of the village social organization here could utilize the high unemployment rate, that is, put some of the jobless people to work regreening what used to be a Green Cape.  The landscape looks like a desert now, but there were thick fields of millet twenty years ago and this peninsula gets more rain than many places with much more green.  When you create agroforestry, the investment is very low compared to building housing or industry and even less than establishing many businesses.  The cost of creating the food and the wood is low, and it has the potential to generate significant wealth.  Once you have this started, then you have the resources to build your village.  I suggest that the people of Senegal – that all of us – consider agroforestry for growing timber and food as an economic basis to help us rebuild our society.

If we can thus raise money and invest it in thoroughly pedestrian ecological city projects, we will have taken a very long step toward an ecocity economy, the only economy sufficiently large and comprehensive enough to change our biospherical tragedy into a beautiful success.

Church interior, Uspenskiy, Russia.

*"Architectural heritage is left for the descendants of the builders as a last witness to history," says Vitali Lepski. And architectural design is the first commitment to a long future, on the part of those who would build any kind of a civilization. Be careful what we design!*

# 7. Appreciation, Preservation, Transformation – Roots and Challenges

Working in West Berkeley twenty years ago, I was part of an interesting coalition. In an area of town being ever more divided into single separated land uses – industry here, housing over there – we were supporting the earlier mix of uses that once existed in this the oldest part of town. We thought it was the wave of the future. In the 1860s through the early years of the 1900s homes were only a few doors from light manufacturing. There were corner stores, a small shoemaking shop, pier and fish market, a lumberyard, a park with Saturday afternoon band and revelers from San Francisco across the bay who came by ferry. For a while – talk about mixed uses! – the bar at night served as the grade school by day. Our coalition was made up of historic preservationists, neighborhood activists, and ecocity advocates with visions of windmills and pedestrian cities in their heads. We represented the past, present and future – and figured there was enough room in town for all three. What united the coalition through four years (though our cause eventually lost to a single-use housing project) was a very broad appreciation. We were the folks with a sense of time.

So it is in this chapter. We move into the future best when we know where in the past we come from. Conversely, becoming more clear on a healthy future adds yet more depth to our appreciation of the past. In addition, being people with a sense of "agency" as Robin Standish called it, we know what we do in the present is responsible to the truth in the past and the best in the future. Thus we look to our roots with the help of a few glimpses from Amalia Cordova of an Ancient Pre-Columbian village in the Atacama Desert of Chile and from Marina Alberti of Venice, Italy. Both these communities have an ancient and enduring life and death relationship with water – each of a very different sort. In the desert, Caspana's tightly-fitted stone irrigation ditches preceded the famous precision aquaducts and walls of the Incas. The Lagoon of Venice was a gigantic moat protecting the first new city of the post-Roman era from the boatless barbarians.

Vitali Lepski introduces us to the ancient towns of Russia, rising through the wheat fields, orchards and snows of one thousand years, and present-day efforts to protect them. Lee Boon-Thong speaks of the traditional villages of the Sarawak rain forests and the influences decimating these communities today. He concludes that these villages should be brought into the development mainstream more gently than is currently the practice. Are there other options, even returning village lands to nature as Mike Vandeman suggests later in these pages? Here we can see development/traditional village dilemmas in full glare.

Then we look to Chile again as Ivan Cartes describes efforts to build for a sustainable future by way of new housing policy in Conception, Chile's second largest city. Then Luis Jugo of Merida, Venezuela describes the beginning of his own town's journey toward becoming an ecocity. With firm roots in the past and clear dreams of the future, we can begin to grasp what must be done.

# Caspana – Pre-Columbian Village in the Atacama Desert of Chile

## AMALIA CORDOVA

The village of Caspana is a settlement of pre-hispanic origin in the Atacama Desert of Chile. It can be viewed as a living projection of traditional culture. The indigenous community of Caspana survives and is very active, organized by firm traditions and deep respect for their ancestry – the ancestory of the Atacamenos people.

The Atacama Desert is located in the north of Chile, and is considered the driest desert in the world. In some places here, the rain has never fallen in all of recorded history. The desert climbs toward the interior, until it meets the Andes. Caspana, at an altitude of 3,362 meters (11,050 feet) above sea level, is one of several traditional villages located in these highlands. The narrow Caspana River canyon carves through volcanic rock, and the village is sited where two tributaries join. The river valleys of this region were first occupied by hunters and gatherers from as early as nine thousand years before the present.

A second wave of inhabitants appears to have settled around 3,000 BC; their main activity was camelid hunting and herding. (The camelids of the Andes are the llama, alpaca, guanaco and vicuña.) Camelids were later used in caravans to transport goods to the Bolivian and Peruvian highlands, as well as the northwest of Argentina. Today, goats and sheep also have been introduced.

With the practice of agriculture, a more permanent form of settlement took place by the rivers. This eventually led to the design of waterways and terraces used by the Incas and other pre-hispanic indigenous populations. This watering system involved strict control and organization, and it is still in use today. Water is shared by the community. It is distributed by turns, down two main canals (north and east).

The main waterway comes from a reservoir located 10 km (6 miles) east of the village, and runs along the canyon. From it, secondary canals distribute water, regulated by sliding doors (stones or metal plate) to the terraces. The canals are made out of tightly fitted stone. Cement is used only in the canal's upper regions, close to the reservoir. Watering is usually done before dawn to avoid evaporation. The plants produced include corn, lettuce, onions, garlic, prickly pears, apples, apricots, pears, many other foods and flowers.

Currently, the community is organized by an assembly of village land and livestock owners, called "comuneros." The population of Caspana is around 480 people, of whom 84 are comuneros. The village has a water committee with one man in charge of the water – the water chief. The position rotates, so every family head must be the water chief at some point. This ensures that everybody gets an equal amount of water.

The waterways are cared for collectively. Once a year, at the beginning of spring, there is a festivity called the "limpia canales," the cleaning of the canals. The whole village participates in the cleaning of the distributing canals, engages in ceremonies and feasts by the waterways along a very long table where everybody eats together. One family must feed the entire village. Past quarrels are amended and the main canal is opened. There are seven other villages in the region, and they all practice the same system.

The terraced land beside the river was originally used exclusively for agriculture. The villagers continue the tradition of periodically burning off the vegetation along the river, so that it will grow up again and be very vigorous. In recent times, houses, corrals and gardens have been built on some of the terraces, but it is still considered farm land. Houses are clustered along walkways and tend to be located toward the higher part of the canyon while the riverbed remains reserved for cultivation.

The oldest structures of the Caspana region are related to herding – isolated temporary dwellings in the desert highland – and agriculture; the first dwellings in the canyon were constructed on a rocky terrace overlooking the river. This part is called "Pueblo Viejo" or old

*Amalia Cordova, Curator of the Museum of Pre-Columbian Art in Santiago, Chile, provided a paper and a talk with gorgeous slides of the spare and spectacular town and setting, people and animals – from which this synopsis was drawn.*

Speakers Amalia Cordova and Elizabeth Leigh with hosts at the beach at Yoff.

village, and its buildings date from 400 AD to the start of this century. Even today, people spend much time up in the highlands herding their animals and come down into the village to water their crops.

Today buildings are constructed in very much the same way as those of the first Atacama Indians. They are mainly made of local materials: walls of fitted volcanic rock with earth mortar, straw roofing, and cactus wood roof beams. The walls are very thick, creating a thermal barrier which resists the extreme temperature oscillation of the desert highland. Heat is obtained by burning wood or "llareta" (azorella compacta), an autocthonous plant now bordering on extinction. Roofs are mostly made from a local fiber and are relatively easy to care for. They provide much better insulation than zinc sheet metal. Windows are small and do not usually have glass; they use shutters instead. New materials such as cement mortar and zinc roofing are used but are not considered as effective or as beautiful as the traditional ones. Unfortunately, they are becoming cheaper or easier to obtain, and generate disposal problems.

Houses are mainly used for sleeping, eating and storage, as most of the time is spent up in the plateau herding animals, out in the "fields" planting or harvesting, or in the village engaging in craft or housework, which is also mainly done outdoors. Some houses have a *ramada*, a roof suported by poles, which provides shade and contact with the open public space.

The village has a church, chapel, several cemeteries at the edge of the village, a school, a museum, community center and two stores. Caspana has no post office, no police station, no gas station, and one telephone located in the school and considered to be for community and public access. Villagers periodically make the trip to a distant larger town mainly for commercial, legal or health matters. There is a private, local bus that runs once a week, leaves one day and comes back two days later. Mail is brought in by whoever goes into town or by the bus.

The community owns a generator (petroleum-fed) but only runs it for a couple of hours each evening. There is a great interest in acquiring solar technology to replace the generator. Inorganic waste disposal is becoming a problem. Inorganic garbage has been produced in the village for only the past twenty years, and dumped at two locations on the opposite edges of town. Before that, waste was not a product of the village.

Caspana's economy depends partially on its arts and crafts. Cactus, wood and stone carving is practiced and sold in the village and in the closest cities and markets. Textiles are another important source of work, employing men, women and children in spinning and weaving and involving women only in embroidery. Traditionally, textiles were made from camelid wool or cotton, with natural dyes. That is mainly true now but some lambs wool and industrially produced yarn is purchased from outside the village and integrated into their work. The Caspaños often use themes rooted in respect for their tradition and their ancestors. One workshop, for example, is producing wool wall hangings inspired by local rock art. Before beginning, the artisans hold a special ceremony asking the "grandparents" for permission to use the images. Another aspect of respect is seen in the 1994 establishment of the "Indigenous Community Center," constituted under a new national law which recognizes indigenous populations as local authorities.

Caspana is a traditional village surviving in the desert and providing a model for other indigenous and alternative communities, while the larger cities of Chile are carrying out a process of intense, consumer-oriented, environmentally unbalanced "development." Communitarian and religious aspects of the Atacameño culture have been fundamental in this process of resistance and evolution, as tradition is deeply rooted and respected by the villagers. Not all Atacama villages have the degree of productivity, relative autonomy and traditional ceremony that Caspana maintains, and these are some of the reasons Caspana is considered a leading village in the indigenous community.

# Venice – Firm Foundations for Future Cities... Built on Water

## MARINA ALBERTI AND EDITORS

Venice teaches us that humanity is able to create the most extraordinary balance between the built and the natural environment, but Venice also reminds us that humanity can bring this delicate balance to collapse.

In Venice the natural environment shapes the urban structure and organization, establishes the limits to growth and sets the rhythms of city life. The city was born on the northwest edge of the Adriatic Sea in the fifth and sixth century. It rose from marshy low islands so flat they barely broke the surface of the waves in its great barrier island-protected lagoon. Why settle in such an unpromising location? Barbarians were picking their teeth with the bones of Rome at about that time, but they were stuck on the land with no knowledge of ship making. In this the first Veneti were knowledgeable. City building was another story, on soft mud. And sustenance and commerce were a challenge with nothing but fish and birds, reeds and a few stands of trees for neighbors.

So the early citizens of Venice learned to drive wooden pilings into the floor of the lagoon and scoop up mud from the bottom, thus creating the first of the 180 islands that now serve as foundations for the city. With perseverance and a simple technology we might call "appropriate" today, they demonstrated that what now is generally regarded as impossible in our world of modern technology – the building of a place as remarkable as Venice – was possible. (They thus raised the question, why can't we summon similar imagination and build ecocities and ecovillages by the end of the 20th century when, presumably, our technology is more advanced?) On their flat islands in sheltered waters they discovered they had a commodity that Constantinople needed in the days before refrigerators: salt, a commodity Venice produced by evaporation.

Jane Jacobs makes a crucial point when she describes this first chapter of Venice history in CITIES AND THE WEALTH OF NATIONS. Venice developed "by acting like Constantinople.... This may seem laughable, that a primitive little settlement of fishermen, salt evaporators and loggers at the back of nowhere could start behaving like rich and mighty Constantinople at the very hub of things, but it did." Venice began copying the production of far-away Constantinople, and over the slow-moving decades, began selling its products not only to its own people but to other later-developing cities around Europe, then all the way to China by the time seven hundred years had passed. Says Jacobs, Venice thus demonstrated how cities develop into the primary economic engines of civilization: "import replacement." They begin manufacturing what they used to import. More specifically and in very physical terms, cities assemble the parts of the engine of prosperity – the resources, the people, the will to produce and trade – all in one small area. This convenient physical access of the crucial ingredients, in the case of Venice, is small enough in land areas as to be 100% pedestrian accessible, assisted only partially by "water taxis" and "water buses."

Today the 180 islands of Venice are connected by four hundred bridges linking exclusively pedestrian streets, buildings, small courtyards and grand squares. Only one bridge connects the city's one modest dead end

*Marina Alberti, who has lived in Venice, Italy and attended the University there, delivered a presentation on the city – a city more than 1,200 years older than its country and a city in 1996, centuries ahead of its automobile dependent contemporaries. Her talk with slides was very short, since she was doing so much work for the conference with another talk and helping with the Conference Proclamation as well. So we have added a few more details in this version based on hers. Therefore it is really a joint authorship: Marina's and the editors.' We added a few words from Jane Jacobs' writing and some information about the successful half of the effort to stop Venice from sinking. The editors can't resist the temptation to point out that there is a strategy that could protect Venice, answering Marina's bleak riddle at the end, though it seems radical indeed: build ecocities and cut off most global warming at its source. In this, Venice, the city without cars, is already doing more than any other to prevent the rise of oceans – if only the rest of the world would notice.*

Bill Mastin

Venice, Italy, a jewel among the world's cities, nervously awaits the greenhouse effect.

parking lot with the mainland. Mostly, the bridge is for the heavy rail traffic to Italy and beyond.

Big trouble was becoming evident about thirty years ago in Venice. From the beginning of this century, the difference in level between land surface and water surface in Venice has changed by 25 centimeters, or almost one foot. Of this, one centimeter is attributable to the rise of the Earth's oceans. In November 1966 Venice was submerged by water for thirteen hours continuously, causing all activities to stop. This produced a call for international attention to protect this world heritage site.

The engineering firm of Gigi Cavaleri was brought in to investigate, and discovered that the main cause of subsidence was the pumping of water from under the nearby coastal industrial city of Mestre. The whole region was sinking with Mestre at the epicenter, though people didn't notice it there since the elevation of the land was originally far higher above sea level than Venice. Officials prohibited further water extraction for industry and instead brought in water by pipeline from the Alps, at which point subsidence ceased. Now Venice waits nervously to see how much further the greenhouse effect might cause the oceans to rise.

On average, flooding now occurs 40 times a year which is six times more than at the beginning of the century. At this rate, there is strong concern that the lagoon could in the future gradually become a branch of the sea unprotected from the larger waves of the Adriatic. During the last 30 years, most attention was given to technical solutions such as building a special system of tide gates at the openings of the land barrier separating the lagoon from the sea. Today however, the urban community in Venice is aware that the problems are much more complex. The delicate balance between the built-up and natural environment is being threatened by an urban economy and style of life that increasingly conflicts with the natural environment.

For example, the water pollution is vast. Since the beginning of this century, organic pollution has tripled and the nutrient load is five times higher. The lagoon receives each year 10,000 and 12,000 tonnes of phosphorous and nitrogen respectively. Aquatic life and fishing have suffered greatly as a result. The solutions studied have had the ring of desperation: releasing artificial floods to periodically flush the lagoon water out to sea, transforming the economy drastically. Still, the sea is expected to rise.

I started saying humanity has been able to create this extraordinary human settlement of historical, architectural and social value. Whether humanity will be able to protect this value is yet to be demonstrated.

# Preserving Old Towns in Russia

Karil Daniels

Vitali Lepski

## VITALI LEPSKI, IOURI MOURZINE

We will speak about our institute's program in accordance with government order. We hope the program will be of interest for those in other countries and will be helpful for solving their problems, too. The program is oriented towards the problems of historic preservation in central Russia. We will try to illustrate our report with slides that may be rather interesting for people who have not visited Russia. Now I will speak from the report of our team.

In the 9th century the ancient Russian state came into existence centered in Kiev, around which all the Slavic tribes became united. In the 10th century, having accepted Christianity as the sole religion, the rulers of Kiev moved into the then unlimited northern lands. By the middle of the 12th century they found that they needed the assistance of the skilled workers of the pagan population and completely captured the whole territory. They followed mostly the river valleys, and they erected towns with churches and fortresses during this movement to the north.

Many small autonomous fiefdoms came into existence, each with its own capital. The feudal lords were engaged in constant warfare with each other and therefore built fortresses, distinct in their architecture, and certain cult structures. As a rule, towns formed in association with the fortresses and later became the historic towns of modern Russia. Many of the Byzantine Christian traditions melded with the elements of local culture. Thus the palaces and churches were built taking into consideration local climate and geography and using local materials and experience.

*Vitali Lepski, Honorary Builder of Russia, is an architect and structural engineer, and General Director of the Central Research and Design Institute for Comprehensive Reconstruction of Historic Towns INRECON. He is a professor at the International Academy of Architecture in Sofia, Bulgaria and here his talk is translated by Iouri Mourzine, Honorary Architect of Russia and Head of Architecture for INRECON.*

At the time when the Russian tribes migrated from the shores of the Dnieper to central Russia, the region was separated from the steppes by thick forests and impenetrable swamps. The fertile Suzdal plain lay in the center of the region; the cities of Rostov, Suzdal and Vladimir were founded on its borders.

In order to increase the cultivated land area, peasants cut down forests and drained swamps. Initially, this took place along river valleys, which provided hay from the low-lying meadows, fishing, and convenient transportation routes. By the 19th century, this territory acquired the look that is characteristic of the Russian national landscape – with immense river vistas, wide rolling plains with mighty forests and hay-producing meadows, fields, near and distant cities, monasteries, and villages with churches on hilltops.

Whereas in the 19th century industrial development and railroad construction gave rise to a new phase in the evolution of the historical landscape, in the 20th century wholesale industrialization undermined the fragile ecological balance. Construction of giant industrial complexes and immense hydroelectric projects, massive draining of swampland, cutting down of forests, the

INRECON

The Novgorod church, built in 1345 in Kovalevo.

INRENCON

Askov Kreml. "As a rule, an ancient Russian city began as a fortress."

changing character of peasant labor, and destruction of churches – all that resulted in a degradation of the historical landscape in vast areas of the country and brought to the brink of extinction a number of species of flora and fauna.

The remaining landscape in the uncorrupted areas is now threatened with neglect and reverting to wilderness as a result of the emptying of village populations into the cities. At the same time, it is precisely because many historical cities and towns lack major industries with their harmful emissions, as well as complex networks of railroads and highways, and because they have tradition-ally, back deep into antiquity, served as foci of emigra-tion, that they have been particularly appropriate for populations displaced by ecological or social upheavals, and as tourist attractions.

Architectural heritage is left for the descendants of the builders as a last witness to history. It is only in the last few decades that human society recognized the historical and aesthetic value of these ancient cities. Other countries also have suffered from tragic short

sighted removal of old buildings in order to build contemporary structures. In addition, in our country during World War II, not only thousands of monuments were destroyed, but also entire cities were wiped from the face of the earth – the history of architecture is not only a record of achievement, but also of tragic loss. Therefore it is particularly important to preserve whatever time has spared. Not everybody shares the preservationist view. For many people, today's problems seem more important than any particular town as a witness to history.

Russia possesses many outstanding architectural monuments. I am not going to dwell on Moscow and St. Petersburg, which have been sufficiently researched. Small historic towns are characterized by everyday homes and civic architecture that organically encompasses churches, monasteries, and civic structures. As a rule, an ancient Russian city began as a fortress circumscribing civic and ecclesiastical buildings. Such architecture was built self-consciously to enhance the local landscape and to organically merge with it. Ancient Russian architec-ture is characterized by a love of ornament and detail

INRECON

Few examples of old wooden architecture remain in Russia. This one is at Kizi.

which tends to be graceful. The interiors of these structures often impress with opulence.

Russia is a country which builds not only in stone but also in wooden architecture. Unfortunately, thousands of old churches have been lost over time as they are very vulnerable to fire. Now days the country has created numerous museums of wooden architecture dedicated to preservation and renovation. In the center of these ancient cities, there was a market square which consisted of rows of shops and workshops. In our definition of an historical city, the system and elements of the original city planning must be preserved: the block, the squares, the historical and cultural monuments – all of which represent the historic whole.

Regarding the natural as well as built environmental aspects of Russia's preservation efforts, there presently exists a partially realized system of ecological protection devoted to the preservation of the disappearing environments and landscapes. These are essential to replenishing the physical and spiritual needs of human beings who are denied access to nature in the ever growing cities. The system includes natural preserves, national parks, green urban zones, recreational zones, protected watersheds and water supplies, and hunting preserves.

The current program of rehabilitating old Russian cities historically, culturally, and ecologically is supported by the government of Russia and is particularly focused on ecological problems. Even as the program acknowledges that moderately-sized urban settlements are in relatively better ecological shape than large cities, it also recognizes that during the past decades smaller urban areas have suffered negative setbacks and proposes measures towards their stabilization and rehabilitation.

There are several long-range goals of the program. To protect the air there are provisions to remove industrial and auto emissions, and particularly harmful projects are being relocated. The impending structural industrial reconstruction aimed at developing traditional industries in mid-sized cities, which are less harmful to the local ecologies, provides for reduced air pollution. To protect water resources the program provides more stringent requirements for sanitary protection of reservoirs and watersheds, especially small rivers and lakes. Regarding land resources, new construction is encouraged on inferior land and on much degraded urban land, intensive cultivation and restoration of green open space for recreation is encouraged. Solid waste disposal sites are to be modernized. A major goal of the rehabilitation program is to help realize existing governmental programs for the protection of the population from the effects of the Chernobyl disaster and for the rehabilitation of the Ural region from radioactive pollution.

The government program has operated in hundreds of cities cleaning up radioactive pollution, retrofitting and reconstructing industries with especially harmful ecological effects, decreasing auto emissions that are particularly harmful to historical and cultural monuments, and preparing plans to protect buildings against flooding due to dam and artificial reservoir construction, and soil erosion. In general the program is dedicated to assuring the viability of traditional settlements and to on-going protection of our cultural and historical legacy for future generations.

The United Nations International Forum of 1992 in Rio de Janeiro established that ecology and socioeconomic development can no longer be viewed as isolated from each other. The "Agenda for the 21st Century," adopted by the Forum, calls upon governments to develop national strategies for stable development. The program of rehabilitation of historical cities and their surrounding landscapes represents one of the steps towards a stable development in central Russia.

In the 1960s through the 1980s, our cities were built primarily utilizing multi-story construction, which resulted in long commutes, as well as isolating people from decision-making and distancing their every-day activities from the remaining nature.

In contrast, the smaller cities and towns are much more responsive to remedial measures aimed at ensuring their survival and stable development. Many of them, for example, do not possess, or possess only in part, a centralized water delivery system; thus they are receptive to the most efficient and state-of-the-art energy and water conservation technologies, and they are the communities most likely to adopt biological methods of waste disposal. As individual auto ownership is not nearly as widespread in Russia as it is in Europe and America, Russian cities, so far, are in a position to avoid the problems that plague towns and cities in other countries, such as great volumes of paved areas and parking lots, harmful emissions, accidents, noise and so on.

Thus the "Program of Resurrection" of Russia's historical towns and cities covers all aspects of their preservation and future cultural and social development. This is not about a conflict between the city and nature; it is about their co-existence. It is imperative that we preserve nature in our cities and welcome it back to where it was previously crowded out. At the same time, we understand that positive results can happen only on the municipal level. Regional initiatives and urban and regional consciousness raising can occur only if there are concrete, positive examples to follow – issuing directives will never achieve the same results. And the most important component is the participation by cities themselves, with the full understanding that the environment must be preserved globally and passed on in a renewed and improved condition, to future generations.

INRECON

Sculptural details exhibit the love of ornament characterized in ancient Russian architecture, as shown in this example from a church in Vladimir.

# Development Influences on Sarawak Villages – The Need for a Paradigm Shift

## LEE BOON-THONG & T. S. BAHRIN

Sarawak, the largest state in Malaysia, has long been considered the most backward. It may be said that the area is inhospitable, with extensive coastal swamps and interior mountainous regions covered with dense tropical rain forests. Rivers, which are plentiful, have always been the main mode of transportation for those living in the interior to travel to and from the lowland coastal towns. The 1.8 million population is ethnically diverse and generally scattered with a major concentration in the coastal urban areas. About two thirds of the population are indigenous and the rest are immigrants. In the interior, about a quarter of the indigenous people are still involved in slash and burn, hunting, fishing and food-gathering. However today, a substantial number have converted to some form of sedentary or semi-sedentary agriculture. Until the early 1970s, internal migration was negligible; few indigenous people left the rural areas for the towns.

After Sarawak joined Malaysia in 1963, more funding was available for development purposes, but lack of infrastructure and the general ruggedness of the terrain still hampered serious development efforts. Up to the 1970s, there was little industrialization, limited large-scale commercial plantations, few roads, and public services which, if available, were focused in the small urban nodes in the coastal areas. Since the late 1970s and especially from the 1980s, the pace of economic development in Sarawak has intensified with increased offshore mining of petroleum and natural gas, logging activities in the interior forested areas, and industrial activities in the lowland areas. Location of major industries was all

within the lowland coastal regions. These have provided employment opportunities for some 30,000 workers. Major infrastructural developments have been made with the building of the Pan-Borneo Highway, ports, and the provision of amenities. These developments in the lowland areas act as magnets for movement of labor from the interior and encourage development in the peripheral interior areas.

From a spatial-geographical point of view, the process of development may be said to involve primarily the interaction between three geographical units: first, the major urban areas, second, the primary circulation system (that is, the network of high-density transaction flows of goods, peoples, and ideas), and third, a mobilized periphery (those areas interacting most intensely and in closest proximity to the urban areas). A fourth geographical area that has yet to interact with the growth and development of the mainstream flow, at least until recently, is the relatively unmobilized periphery. This is the interior upland region where poor communications and difficulties in transportation have kept the area relatively immune to the development that is taking place in the lowland primary circulation system. A gradual transformation is, however, slowly taking place in this unmobilized peripheral area as innovations, values, and ideas have begun to penetrate into the interior hinterland.

It is possible to put forward three models of settlement change among the traditional villages in this area of Sarawak. Model A is a situation in which the currents of development have little effect on the traditional communities by virtue of their extreme isolation. These villages remain well-integrated social units because very little outmigration has occurred. Models B and C are the transition models which we will address here with actual case studies. In Model B, the diffusion of innovations into settlements has brought about gradual adaptive change in the structure of the settlements. Changes are accepted proactively; for instance, the use of cylinder gas instead of firewood for cooking. In Model C, the tradi-

*Lee Boon-Thong is a professor in the Department of Geography, University of Malaysia in Kuala Lumpur. He presented a slide show taking us up the rivers of Sarawak and into the remote traditional villages of its rainforest, featuring the traditional tribal "long houses" and farming methods and cultural traditions of the people. The paper he submitted, from which parts of this version were drawn, was co-authored by Tengku Shamsul Bahrin.*

tional settlement faces the sudden onslaught of development forces from outside, such as logging activities and land settlement schemes, and changes occur at a much faster rate than in Model B. These forces normally tend to increase geographical accessibility through the construction of roads. Both settlement types B and C have yet to be incorporated into the larger system of development by virtue of their relatively inaccessible locations.

The results of both processes, whether a gradual diffusion or the intrusion of exogenous forces, are more or less the same. Young adults have begun to migrate to the lowland areas in search of further education or employment opportunities, leaving behind the older folks to tend to the fields. The sociological impact is, of course, a gradual disintegration of patterned ways of acting among the people who share some common sentiments. Village life changes, with an increasing value placed on monetization and materialistic accumulation. Television and audio equipment of those who stayed behind have transformed the once-tranquil villages. The older folks bemoan the dearth of life in the villages compared to yesteryears.

The Baram villages provide a model of rapid change. The 440 km-long (375 mile) Batang Baram is one of the main rivers in Sarawak, and along tributary rivers located about 150 km (93 miles) from its estuary are found pockets of native Kayan settlements. The Kayans, the second largest indigenous group in Sarawak, are a closely-knit riverine longhouse community. Initially involved in fishing, hunting and some shifting cultivation, they eventually became semi-permanent agriculturalists growing hill padi, coffee, rubber and subsistence crops such as maize, vegetables and tapioca. Although the Kayans had contacts with the outside world through trading their produce, they were still fairly secluded from external influences that would disrupt or change their lifestyles.

Christian missionaries who came after the war succeeded in converting entire Kayan communities. Pagan practices and superstitions gave way to a village life centered around a church whose activities further consolidated social interaction. Education was also introduced at about the same time, and a small number who passed their primary examinations eventually filtered down to the lowlands to continue their studies.

In the early 1980s, the bulldozers came into the forests. As the timber industry penetrated deeper and deeper into the interiors, the effects on the Kayan settlements were basically threefold. Firstly, timber tracks increased the geographical mobility of the Kayans. Settlements began to be realigned to land transportation instead of the less efficient river transportation. Secondly, settlements received money or amenities (such as electricity) as compensation for intrusion into communal land. Thirdly, locals were offered attractive wages by the industry. The overall result is that the Kayans became more aware of the outside world and were able to move more easily into the lowland areas – at a rate twice that of outmigration from small towns in Peninsular Malaysia. About one-third of the migrants moved within the Baran "outback" as logging companies made jobs available. Another two-thirds moved to the lowland towns either to resume further studies or to follow their spouses or to seek employment opportunities. More than two-thirds were between 16 and 35 years of age, underlining the loss of able-bodied labor from the villages.

The village of Bario provides us with a model of more gradual change. Bario is the main settlement of the Kelabits, one of the minority indigenous groups in Sarawak. Located in the Kelabit Highlands about 1,000 meters (3,300 feet) above sea level, it is deep in the interior of Sarawak beyond the farthest reaches of the navigable rivers of the Baram District. There is no road to Bario. One must be prepared for a week's walk through the forest and highlands to reach Bario. Consequently, prior to the Second World War, the entire Kelabit population of 1,600 was found in the Highlands. In fact, historically, the Kelabits have been isolated from other native groups for more than 500 years.

During the Second World War, external influences began to penetrate in the form of Christianity and guerilla bases against the Japanese. Gradually, educational and medical facilities were introduced. In 1953, missionaries opened up an airstrip and in the early 1960s British soldiers were stationed in Bario which brought candles, movies, bread and canned food.

In 1980, the total Kelabit population in Sarawak stood at 3,800 and this was estimated to have increased to about 5,300 in 1992. However, as the area became more and more exposed to outside influences, many of the inhabitants had migrated to the lowland areas in search of jobs and further education. Robert Lian Saging and Lucy Bulan estimated in their 1989 report to the Sarawak Museum Journal that more than 50% of the Kelabits had migrated to the lowland urban areas since the 1970s.

A survey I conducted with Tengku Bahrin in 1993 found an average of 3.9 persons per migrant household – a rate very much higher than the three persons per migrant household of the Kayans in the Baram River basin. The bulk of the migrants (85%) had moved away from the highlands to the lowland urban areas. This shows that lowland development has a substantial impact upon the highland dwellers. The urban-ward migration is largely dominated by males who had moved because of job opportunities (70% were males). Contrasted with the Kayans, where outmigration became substantial only after the impact of logging activities in the early 1980s, outmovements from Bario were of increasing significance after the early 1970s, before logging had reached the area. Thus, outmigration may be viewed there as largely the result of increased education. In the early 1960s, school enrollment began to increase, and by the early 1970s a greater number of Kelabit students began to continue their studies or search for jobs in urban centers in the lowland areas.

It also appears that the trend of migration is unidirectional and of a permanent nature. Only 17% of the total respondents indicated they would go back to Bario to work in the farms. In reality, only one migrant re-ported returning to work in Bario because of the inability to find a job in the urban area. Similar to the case of the Kayans, return migration is not expected because of the higher remuneration in the destination areas.

These implications necessitate a paradigm shift. It is obvious from a comparison of both the Baram and Bario illustrations that settlements cannot continue in their traditional mold given the momentum of development in Sarawak. Regardless of whether there are active logging activities in the interior or not, change is happening. The nature of the migration stream has far-reaching implications for the rural settlement matrix. Conceptually, these areas represent the hitherto unmobilized periphery. But today, improved high powered boat transportation, air connections, and new logging tracks spreading out like tentacles into the remote interiors have caused the traditional population to become more fluid. The consequence is a slow but sure breakup of self-sufficiency and isolation.

The ecological landscape of the interiors is also changing with the encroachment of logging activities. Rivers become polluted and fisheries become depleted. The traditional mode of hunting and food gathering can no longer be sustained as animals become scarce. Seeking jobs elsewhere is not the only alternative left but it is more lucrative, especially among the younger generation who show a growing disinclination to work on the farms. Education and the work of Christian missionaries have also contributed to the growing aspirations of the local people. Thus, a selective migration process involving the more energetic, more motivated, and better qualified drains the traditional settlements of "elitist" elements and leaves behind an aging demographic structure. Consequently, productivity and economic surpluses decline. There is no way in which a traditional society can be shielded from the onslaught of modernization. There is also no justification to stop longhouse-to-urban migration if the people choose to do so.

However, if the economic reasons for outmigration far outweigh others, then it would appear that a paradigm

Lee Boon-Thong

In Sarawak, Malaysia, river travel has been eclipsed by land transportataion, as timber roads penetrate deeper into the interior.

shift in development strategies would be necessary to maintain the interior populations. In other words, opportunities for effective utilization of labor resources should be created to bring them into the mainstream of development. In this respect, land development schemes and an intensification of the rural growth center strategy should be realized, especially in priority and potential areas. The latter strategy should focus on the development and diversification of the rural economic base through increased commercial activities in production, processing and services, besides providing better basic infrastructure and social amenities. Usually, these efforts have been viewed as attempts to impose change from above and attempting to destroy traditional rights and privileges. To overcome these suspicions, "modern" development techniques should attempt to understand local sensitivities and local knowledge. For instance, native customary land tenure assumes that land rights are not to be alienated outside the community. But because the Land Code contains limited provision for the protection of customary rights, land has been alienated for a number of development purposes. This caused insecurities, dissatisfaction and mistrust. On the other hand,

local groups should respond proactively to developmental efforts. Writers M. Cleary and P. Eaton predict the pull of the primary circulation system will give rise to considerable problems in destination areas in future. Creating opportunities and choices based upon this perhaps desperate proactive strategy appear to be the logical paradigm shift to ward off these problems.

In concluding, we have noted that traditional settlements frequently come under pressure with economic change and the transition from subsistence to a monetary economy. This process is accelerated when there are active external impulses such as logging. However, even without such exogenous factors, it is not possible to shield villages from the tides of rapid economic development. Depopulation is inevitable because the isolation of these interior areas makes circular mobility impossible. Thus, a more effective utilization of labor within the interior is needed. A paradigm shift in development strategies to create opportunities and choices within the rural interior may help to alleviate depopulation problems even though the efficacy and results of such a policy are, as yet, uncertain.

# Readdressing Planning for Sustainable Development in Chile

## IVAN CARTES, J.C. CHILTON AND E.R. SCOFFHAM

This paper attempts to demonstrate how Concepcion, the third largest city of Chile, has tried to control urban growth through the implementation of several urban regeneration programs.

In the Third World, a lack of resources has limited the ability to focus on long-range goals in the urban development process. Instead, focus is increasingly placed on emergency and immediate problems which demand smaller, short-term solutions. Thus the process of urban development has shifted toward including more small-scale, neighborhood and community efforts and less macro schemes of investments which cannot fulfill immediate needs as easily.

During the next millennium these heterogeneous social demands and short-term goals will compose, in part, the urban panorama. In addition to defining cities according to the global market, it will be necessary to pay attention to the localities where neighborhood social factors and interests will determine the larger urban context.

Social participation, as a more balanced approach between social needs and final solutions, seems to be quite logical in democracy, but Chile only restored its democratic process in 1989 after a 17 year period of military rule. In that period, planning processes and urban design schemes were used as a means of controlling the population and their subsequent settlements. For example, squatters were not well received within or close to the city's edges and, on several occasions, the people were cleared from the capital to other less populated urban centers in the country.

The idea of having tidy cities directed the planning processes towards pure physical improvement rather than consideration of the inhabitants' memories, social liaisons, and sense of belonging to the city. Indeed, these characteristics constitute key factors for integrating the inhabitants in the process of development, because this is the particular place in which they wish to live and the one that they want to enhance. What is needed is a more supportive attitude which, instead of denying their existence, searches for ways of supporting their initiative and improving their existing conditions.

Since 1989, the last two democratic governments have consciously worked towards restoring the people's confidence, making them aware of urban affairs and stimulating their local participation. The recent Law of Municipalities approved by the Congress empowers the communities and neighborhoods, thus bridging an important gap in decision making. This new step towards the democratization of the city has produced a more open administration in local councils, whose elected representatives now have to prioritize urban development issues according to the support or disagreement of their inhabitants. In this sense, the most important guarantee of success is popular consensus and approval. This process proposes a more balanced approach for urban development, thus reducing social conflict and creating a more equitable distribution of opportunities in the cities. If the inhabitants understand the development process and take part in the creation and management of their own habitat, a sustainable urban system is guaranteed.

Environmental changes strongly influence the quality of urban life. People's behavior changes as they change their environment. This principle is essential to the creation of city development that improves both the environment and quality of life.

The notion of a "system" should be introduced whenever an opportunity occurs at the community level. At this time the local government can explain the development alternatives and their future implications. In vain, local administrations often try to force development in only one direction. However, if the community is

*Ivan Adolfo Cartes almost made it to Senegal, but the government there lost his visa application until the very day his airline payment deadline came due. He was very unhappy but wanted his paper, co-authored by Dr. J. C. Chilton and Dr. E. R. Scoffham, made available nonetheless. Mr. Cartes is an architect and urban planner from Concepcion, Chile.*

shown potential solutions or a series of choices, they will support and work hard for whatever appears to be the best or the most appropriate solution.

Whenever urban contentions arise, either by the community or by the local administration, the problem and possible solutions must be examined by showing all the interrelated factors and implications concerning that conflict. For example, innovative schemes of traffic calming have improved the air quality, which in turn has created higher standards of health, and consequently the idea of a better quality of life. A small action or solution does not necessarily imply a narrow focus. What is needed is an interdisciplinary focus and the informing of the inhabitants about the matters under consideration, addressing environmental improvement as an interrelated process. Partial solutions are always a waste of time and money.

The country of Chile is well known for its current economic stability but less well known for its initiatives combating poverty and providing more opportunities in urban areas. It is considered a wealthy nation and has been internationally recognized as one of the most dynamic developing countries. However, despite this symbolic change, it is worth mentioning that the usual exclusion of low-income families from modernization and opportunity was prevented through the design and implementation of a more equitable economic system. In order to ensure a sustainable form of development, then, a more egalitarian perspective of economic distribution is required. Said the World Bank in its 1995 report "Preparing an Agenda for Habitat II, Istanbul," "specific form adopted by economic growth also defines the distribution of power; that is, the capacity for action by different social sectors to maintain or increase their quota of participation in society."

The last two democratic governments have made a great effort toward more inclusive social and economic policies. They turned their attention to the poorer sectors and to the integration of "micro-enterprises" within the national prosperity. This initiative also oriented production towards international markets, thus contributing to national exports. A good example of this is the program run by FOSIS (Fund for Social Solidarity), which helped 1.5 million small factories, each employing five people or less, to achieve quality high enough for export. This opened a new era for the so called "non-traditional products for export." Several of these small factories are making products from regional or recycled materials, with the intention of adding an extra value to the final products .

Other initiatives, for instance, allowed systems of water supply to be built by constructing small reservoirs or ponds, filled with natural sources of water and delivered to the same area. This decentralization generated economic power that has brought self-confidence to these communities. To support these programs, the tax system was adjusted to stimulate the sectors where the need was greatest. These social programs reflect the acceptance that the poor have an important role in the country's economy.

By the end of 1990, the proportion of the poor in the population had dropped below 15%, demonstrating that by tackling disparities the idea of growth and equity can become a reality for a developing country. These achievements have demonstrated that the economic system with an equitable distribution of opportunities and wealth can also reduce social conflict in the move towards more sustainable urban development.

In the Third World it is impossible to continue with the notion that the Government is the only agency capable of providing housing or meeting everyone's needs. It is true that the central government has to look after the general process of national development, but with the growing problems in our cities, it is better to stimulate mixed initiatives or the enormous task of solving the country's problems will never be completed.

This is why the development of a "public-private partnership" is favored as an alternative and neighborhood regeneration as a site for focused efforts. These are small or large enterprises, which bridge local residents, the private sector and/or the council. This kind of association balances multiple interests from different sectors and fosters a general agreement among them. This is the base for any equitable and sustainable development and its structure automatically regulates actions in the present as well as in the future.

In Concepcion, several important programs were implemented in order to enhance the quality of housing and neighborhoods in the city. Strategic and meaningful improvements were done by the "squatter enhancement program" and the "progressive housing scheme" on existing sites. These have elevated the standards of health, provided services and supported community action with self-help methods of construction. Other schemes were carried out to improve inner city areas, such as the water-front development beside the river and the renewal and regeneration of other run-down neighborhoods.

These programs changed the citizen's perception of the city administration and gave citizens a much more positive understanding of the urban development process as whole. Several governmental institutions have promoted a series of "partnerships" negotiating mixed economic and social interests, where the notion of consensus has attracted further initiatives and businesses.

Over this period of time, art in the city has come to constitute a central theme for collective action and expression in urban communities. In fact, the philosophy of "Planning for Real and Urban Action Groups" is based on the concept of expressing popular needs through many different types of artistic endeavors.

Art in the community is also closely related to leisure and recreation. Once people comprehend that art can be a method of fulfilling their recreational needs within a community, they are more willing to participate. In this sense, the concept of "amenity" in the urban space is critical for stimulating people's pride of place.

In Concepcion, a major project of art and urban design was launched by the local University and the City Council in order to enhance public spaces in several areas of the city. The plan involved cooperation and action between several disciplines. Working in run-down neighborhoods, sculptures were made by local artists and built with regional materials. These adopted areas gave the neighborhoods a new sense of place by integrating art, urban design and community participation. It also has provided popular leisure venues for people of all ages.

All these initiatives – involving citizen participation, supporting ecologically healthy development, combating poverty, encouraging art – have created a positive perception of the city, improved the quality of urban life, and increased concern for the environment. They have also demonstrated that a holistic approach works much better as a general process of urban development in our cities and provides more balanced solutions.

However, institutional practice is still far from ideal solutions because the public sector does not tolerate or acknowledge mistakes in the urban development process. But it is mistakes and feedback which are crucial. In the developing countries, environmental development needs to progress in such a way that both achievements and mistakes can be positively incorporated into the system.

In this sense, it is difficult to teach communities – especially those with long-held traditions and different backgrounds – which environmental principles they should follow. It is important to present environmental issues in an interdisciplinary context and inform the inhabitants about the choices and possible future implications of each. Then the inhabitants will begin to feel that the environmental improvement in the urban context is a direct result of decisions they have made through their own urban development plans.

Thus, social participation is fundamental in this more conscious process of urban development. That is to say, an equitable and sustainable urban development will only become reality if the people identify with the development of their city. The challenge of making economic and urban growth consistent with an equitable distribution of opportunity and welfare continues to be extremely important in the Third World. Consequently,

Richard Register

Along the western flank of the great Andes mountains, air pollution gathers from automobiles and industry. In this aerial photo, smoke collects in the lowlands between Santiago and Concepcion. From the ground, the 22,000 foot peaks are obscured.

urban development requires that economy, social welfare and environmental quality advance simultaneously.

As the notion of a paternalistic central or local government doing absolutely everything becomes obsolete, "partnership" will become increasingly more important. Current approaches to urban development have demonstrated that, especially in the areas where mixed tenure and multi-sector initiatives have been implemented, partnership has produced better results for regenerating neighborhoods and reviving the social patch-work and the subsequent urban fabric. The development of "public-private partnership" appeared as an alternative by facilitating co-operation between local residents, the private sector and/or the council. This approach balances multiple interests and emphasizes the general agreement for a more equitable urban process, which is the basis for a sustainable development.

# Mérida and the Albarregas River in the Venezuelan Andes – A Potential Ecocity Project

## LUIS JUGO

Mérida, with 250,000 inhabitants, is a sinuously winding linear city in a valley of the Venezuelan Andes. It is identified as a University City because it is the headquarters of the bi-centennial Universidad de Los Andes, one of the most important schools in the country with more than 30,000 students.

The Albarregas, on which Mérida was founded, is a small river in a sub-basin of the Chama river, rising at more than 4000 meters (13,000 feet) above sea level, descending abruptly to 1900 meters (6,300 feet), where it enters the metropolitan area of the city of Mérida. From there, it runs softly over 13 kilometers, receiving the waters from multiple small rivers in their respective micro-basins. Farther downstream, at 1100 meters elevation (3,600 feet) it flows into the Chama river in the basin of the Maracaibo Lake, at the west side of the country.

Today it has become almost commonplace to hear people advocating that when a river goes across a city, it should establish the axis of a local sustainable development and quality of life strategy. Unfortunately, that's after the fact in regard to most already-built cities. Albarregas river, for example, from the 1940s on, has been converted into an open sewer, receiving, without treatment, most of the sewage of the city.

Since the 1960s, the recreational and landscaping potential of the Albarregas valley for the city of Mérida has been recognized. In the 1970s, by municipal and national resolutions, the river banks were decreed a Metropolitan Park. In the early 1980s, a landscape architecture plan was developed – but never implemented. The improvement of our water courses since then has been small; the plan remains a great challenge for the city's people and institutions as we approach the 21st century.

Today, it is claimed as a global social-environmental project that focuses on environmental sanitation for the river and preserving the ecological potential of the city and its hinterlands. Implementing the plan would help establish an educational strategy, providing the raw material – a real, vital river in an urban setting designed to relate to it – for a global environmental education strategy. Mérida citizens and people from around the world then have a place and an institutional framework challenging academics, students and citizen planners to pursue inter-institutional, inter-disciplinary and trans-disciplinary work. The result could be a similar kind of partnership suggested by Lee Boon-Thong, in our case between politicians, university people and the citizens of the community and the world.

At the local level, part of the proposal is to establish natural areas and botanical gardens for all kinds of students and all accessible by ecological footpaths that connect with Merida's schools and neighborhoods. There would also be recreation, eco-tourism, environmental education facilities provided. As part of the scheme, treatment of sewage by bio-digesters would produce biogas for the park guardian's house and organic fertilizer for both botanical and agricultural nurseries. Organic composting and earthworm composting of organic garbage is part of the plan, along with alternative energy technologies and many other programs for quality of life and local sustainable development.

In this last period of the 20th century, the global challenge for mankind is to promote and to consolidate sustainable development as the principal legacy to our future generations. But the methods and the experiences for this new paradigm have to be proved at the local level.

Each basin – with its sub-basins and micro-basins – requires it own integral management plan focused on quality of life, exploring and using biodiversity without its degradation and eradicating the damage done to the environment in each and every human settlement.

The general guidelines for this proposal are designed to appraise the ecological potential of the human settlements and their hinterlands within both the municipal

*Luis Jugo is a professor of architecture at the University of the Andes, Mérida, Venezuela.*

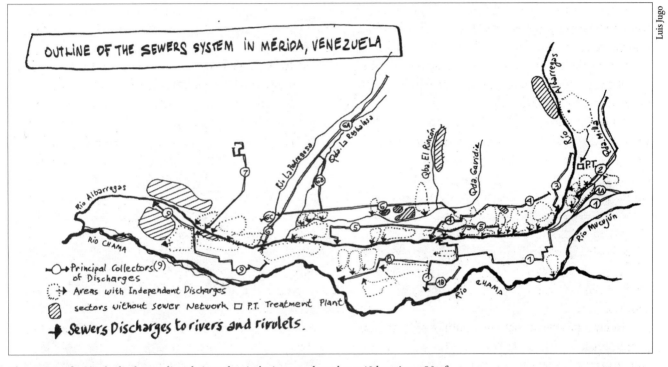

Luis Jugo

The sewers of Méreda discharge directly into the city's rivers and creeks at 49 locations, 30 of which are on the Alberregas river and 10 of which are on small tributaries of the Alberregas.

boundary and the basin limits, since a whole watershed can embrace more than one municipality, establish basin management plans against erosion, define municipal and regional forests, pursue rational harvesting, create non-contaminating alternative energy systems such as solar, wind and water systems, and attack poverty through self-management communities focused on the quality of local life. In addition, the guidelines would help connect civic urban spaces, such as squares and parks, with natural features by way of ecological footpaths, establish river banks as ecological protection and recreational areas and divulge, throughout schools, the techniques of experimental crop production for food, medicine and ornamental plants. These could then be practiced at individual homes, helping the families' nutritional security. In this plan, the schools would also host programs to take care of the ecological footpaths, to engage in forestation

(silviculture), fight against erosion and prepare emergency response to natural disasters. Finally, one of the most important elements of this plan is its proposals for discouraging private vehicles and fortifying urban public transportation networks.

Hundterwasser Haus, Vienna, Austria.

*This intriguing building was built as low-to-moderate income housing by the artist Friedensreich Hundertwasser and the City of Vienna. His cheerful colors and flowing forms attract tourists from around the world. From the social and ecological perspective, the building is a triumph of mixed uses, terracing and biodiversity – with rooftop trees, bushes and vines changing colors seasonally. A small corner store, nursery school, cafe, tourist shop, dentist office, shared semi-private terraces and other uses are all accommodated here. Hundertwasser Haus is like a small village in the city.*

# 8. Ecovillages – Housing and Whole Communities

We have to start small, largely because none of us working for an ecologically-based community has the resources to start large. If we don't start small in the sense of dealing with a small population community, we have to start small in the sense of taking on a small part of a larger community or a small part of a policy or strategy for change. Again, it's lack of resources.

But there are some things about the village scale for this new experiment in building and living that are not options imposed by limits, but that are actually rather ideal. At this small scale, for example, we can know the players well enough to have some confidence in how they will perform, which is valuable when whole lives and livelihoods and whatever savings we can muster are committed to such work. As for contact with and relating to nature, the village, if not overwhelmed by sprawl, is surrounded by nature, embedded in it and not isolated from it, as are most people in urban centers and suburban neighborhoods. And, as Ross Jackson, Chairman of Gaia Trust of Denmark says, "villages are the smallest scale large enough to encompass all the essential elements of the human habitat in need of transformation." Here we can see what the parts are and how they should connect. At that small scale, we can be ecocity pioneers on our own initiative with little support and with minimal opposition of the sort that larger projects, whether conscientious or unconscionable, almost always inspire. Quiet obscurity is possible, or if the enterprise proves relevant to the larger society and everyone's dreams of a better home, if you want it, publicity too.

Birgitta Mekibes tells us here about the Swedish government's idea of ecovillages: residences with energy-saving features, solar orientation and technologies, healthy building materials, gardens, composting and a certain degree of cooperation. Jorge Sainz Casallas describes the confluence of revolutionary ecological and social/economic thinking that led to the first real ecovillage tourist destination, some will think ironically, set in Communist Cuba. Their more isolated ecovillage, with less sophistocated technology than the Swedish counterpart, supports all activities necessary for community life – "full featured," as Robert Gilman would say. In addition, the village is self-consciously structured around forest and soil restoration work and showcases an ecology center with training facilities and a small hotel. Liz Walker relates the tale of EcoVillage at Ithaca. And Ross Jackson proposes establishing a fund to contribute $100 million to the ecovillage movement's already committed and active communities.

The village "seed," that gave rise to the city 9,000 years ago at Çatal Höyük in Turkey can be planted again, but this time the seed is the ecovillage, and it contains the DNA of the ecocity, as well as its own genetic encoding, fit for a healthy environment.

# Two Eco Villages and their Users: Kesta and Kloster in Sweden

## BIRGITTA MEKIBES

It is noteworthy that Sweden as a nation has committed itself to becoming a sustainable society. At the United Nations Earth Summit in Rio de Janeiro in 1992, Sweden, along with 180 other nations, signed "Agenda 21," a long term programme for action against poverty and other threats to the environment as we move into the 21st century. Agenda 21 is directed toward nations, organizations and individuals, and stresses the importance of having each municipality working for sustainable development of society.

In addition, Sweden has passed a number of laws promoting sustainability, specifically the Ecocycle Bill in 1993 – which mandates recycling of consumed resources as new usable resources – and the 1993 amendment to the Swedish Plan and Building Law of 1987 – which calls for the development of a living environment that is appropriate and sustainable on a long-term basis.

The Swedish government also ruled that by the end of 1996 each municipality should adopt a local Agenda 21. Thus, municipalities in Sweden are required to develop thoughts and action plans that correspond with the laws of nature. Under what is called the "Sodertalje Project," each municipality is to build up local knowledge and to define problem areas as well as trying to develop new planning solutions. The project, named after a town 60 km (37 miles) south of Stockholm, tried to answer the question, "What does municipal recycling planning actually mean?"

The greatest challenge to the success of the Sodertalje Project is to develop environmental awareness within each department of each municipal administration, since these departments all have different roles in the quest for environmental improvement. To overcome this challenge, five unemployed architects spent the year 1993-1994 interviewing the heads of municipal departments, attending lectures and conferences and conducting study trips to special projects in Sweden and Denmark. Through this program we have been looking at such important questions as: What land is appropriate to use for housing? What technologies are appropriate for the provision of water, energy, transportation and sewage treatment? What are the most recyclable building materials? What are the best property management techniques to ensure recycling? What is the best way of increasing ecological consciousness within the housing industry? How can sustainable housing be provided for everyone?

The idea for constructing ecological villages in Sweden began in the 1970s, partly as a result of the global petrol crisis in 1973. But it was not until 1984 that the first Eco Village, Tuggelite, close to Karlstad, was built. In the late 1980s and early 1990s the construction of several new Eco Villages was begun, most significantly Kesta and Kloster. Although these villages do not provide a complete solution to urban ecological problems, they do provide a first step in a more suitable direction.

Kesta Eco Village was constructed by a major building company for the Swedish Housing Exhibition in 1990, and it represented one of the main attractions of the show. In the northern section of the village stood a forest whose wood was used for heating. The village also featured a separating toilet system that allowed farmers to use collected urine as fertilizer.

Kloster Eco Village, developed in 1991, was based on the ideas of a group of neighbors. Although not everything they wanted could be realized, the houses do fulfill many of their ecological aspirations.

However, both Kesta and Kloster have significant shortcomings as environmental communities. In Kloster, most families need two cars in order to manage their everyday activities, thus eating up any gain in housing ecology through a loss in transportation efficiency. Kloster also needs a local school, local businesses, a biodynamic farm and more residents. Kesta's failings are primarily administrative and political. As one owner-resident explained, "Somebody built the village for us,

*From a talk and paper by Birgitta Mekibes, Assistant Professor, School of Architecture, Royal Institute of Technology (KTH), Stockholm, Sweden.*

Courtesy of Birgitta Mekibes

Kesta Eco Village with roofing work in progress. At this residential development, all buildings with their attached solar greenhouses are oriented toward the sun.

then it was supposed to function. The 28 families from different backgrounds and with different aspirations were expected to cooperate in the administration of the Eco Village.

Although there is no consensus as to the definition of an Eco Village, the Swedish National Board of Housing, Building and Planning (Boverket) has tried to establish a preliminary list of attributes. The settlement, according to the board, should consist of a closed local system recycling all wastes. This involves cultivation, food storage and on-site drinking water. Specifically, each household should have access to a piece of land for farming, each house should have a non-energy-consuming room, such as an "earth cellar" for storing food, and, if subsoil water is drinkable, there should be a well. The board believes an Eco Village should separate waste water (from toilets) from "gray water" (from bathtubs, kitchen sinks and washing machines) which needs little or no treatment before being used on land, and complete processing of this water locally.

Concerning solar energy, heating and electricity, the board says Eco Village homes should make the maximum use of passive solar energy and should not be constructed on northeast-, north or northwest-facing slopes, they should consume no more than 75 kilowatt-hours per square meter per year, with hot water being derived from renewable energy sources, and that the electric equipment of the houses should be "sustainable" and conservative in energy use. In addition, houses of such villages should be protected from wind and the exchange of fresh air in the dwellings should be assured by natural ventilation or by ventilation controlled according to needs.

Regarding building material, they should be consumer labeled and should not represent a health risk to the occupants. Communal life should be promoted too, and there should be five square meters of common space for every dwelling.

None of the Eco Villages in Sweden fulfill all these Boverket criteria and, in fact, the criteria themselves are far from universally accepted. Perhaps most strikingly absent from the standards is the omission of the human dimension. However good the physical plan of a village may be, its final ecological contribution cannot be guaranteed without taking into consideration the interaction with its citizens. After all, environmental problems are caused by misalignment between society and nature; future societal processes must conform much more closely to those of nature.

In Sweden we are only beginning to become conscious of the tremendous negative environmental impact of the construction and demolition industries. We have a long way to go to catch up with Denmark: in 1992, in Copenhagen, 80 percent of all construction and demolition waste was recycled, reducing the need for virgin materials in that city by 18 percent.

The three keys to reducing the environmental impact of building and demolishing structures are l.) selective demolition, 2.) on-site storage of materials and 3.) recycling-friendly construction, that is, construction which uses materials and methods that facilitate recycling, and designing building products so that they readily incorporate recycled materials.

Actions to facilitate these goals include: prohibiting the disposal of recyclable building materials, approving

Birgitta Mekibes

Inspired by water movement in streams, British sculptor John Wilkes developed the virbela flowform cascade. This one has been used since the 1970s to clean sewage water at the Steiner seminar center in Jarna, Sodertalje, Sweden.

demolition requests only if a recycling plan accompanies the application, requiring producers to accept long-term responsibility for the impact of dangerous products and setting minimum requirements for the use of recycled materials in building projects. Lastly, it must be emphasized that building materials and waste should be considered in the broader legal, economic and technical context.

In addition to planning for the ecological construction and demolition of structures, there are also ecological issues related to the ongoing management and maintenance of buildings. These include the use of environmentally safe cleaning products, installing superior technology at time of replacement, such as water conserving spigots, installing solar panels at the time of roof repair, sorting trash to improve recycling and creating and utilizing compost piles.

Promising measures for more sustainable property management already exist in Sweden, mainly within some municipal housing companies. Through research and development – such as the work going on at my university – the process could be strengthened. It is essential that theoretical solutions be translated into practical uses, but these will only succeed if carried out in close collaboration with interested property managers.

In conclusion, planners and builders will have to learn more about the ecological approach, ecocycle thinking, sustainable development, recycling, resource conservation and a global perspective. We have to develop methods through which we can use this new knowledge. Planners and architects need a vision where environmental awareness exists at all levels. We see Agenda 21 as a basis for international collaboration, especially with developing countries. The municipality can be the object of the study, and the work toward a local Agenda 21 can be the vehicle. The Eco Villages have broken into new areas and have given us practical examples of solutions as well as of unanticipated consequences to learn from.

The time to merely "think green" is over. It's time now to make real environmental improvement.

# Las Terrazas Ecovillage, Cuba – Historic Ecotourist Destination

## JORGE J. SAINZ CASALLAS

What may be the world's first true ecotourism resort is nestled into the foothills of Cuba's Sierra Del Rosario mountain range, an hour west of Havana. If it is surprising a small, remote village in a socialist country should produce a pioneering ecology center and a whole community hosting ecotourism, it is a story told in political revolution and ecological pragmatism – and an inspiration for aspiring ecological communities everywhere.

The history of Las Terrazas (the Terraces, in English) began with the anti-slavery revolution in Santo Domingo toward the end of the 18th century. French landowners, thrown out of what is now Haiti, moved to Cuba and settled in the region of Las Terrazas, stripped most of the trees from the hills, planted coffee and came close to re-creating their slavery-based economic system. It was called Cañada del Infierno, Canyon of Hell, for the harshness of man and nature, and the isolation and poverty of the place. Now the dense forest canopy, traversed by the San Juan River, hides the remains of the San Pedro and Santa Catalina coffee plantations.

One hundred and fifty years later, in the early 1960s, young Fidel Castro and the other revolutionary leaders' objective was to see to it that the revolution reached the most remote corners of the country, where the quality of life had changed little for centuries and people were just eking out a bare living – largely unaware of what was going on in the world outside.

The idea was to convince these small peasant farmers (campesinos), who usually lived in dirt-floor, thatched-roof "bohios" with no electricity, plumbing or other conveniences and no schools or medical services anywhere near them, to move closer together, to form small villages or farm coops, so that all these services could be brought in. This was harder than it might sound since the first Agrarian Reform instituted by the Revolution had just given the campesinos title to the land they worked on, and it wasn't easy to convince them that they would be better off leaving it or pooling it.

*Based on reports to Ecocity 3 by Jorge J. Sainz Casallas.*

La Terrazas Ecovillage, in a rare combination of mountain rainforest and grassy savanna, is both an ecotourism model and destination.

At the same time, provincial leaders were asked to pinpoint what the main problems were in their areas. Cuba was and is primarily an agricultural country, and so, the most frequent problems were said to be lack of irrigation water and, related to the earlier deforestation, soil erosion and other problems.

In the Sierra del Rosario area, these two factors came together to lead to the adoption of the terrace plan. To reverse the destruction of the forest and soil, the provincial leaders proposed terracing the mountainsides and engaging in major reforestation efforts along these terraces. (The Sierra del Rosario is mainly low foothills, 500 to 2,000 feet high, not steep rocky mountains.) The terraces would help stop the erosion and at the same time begin the restoration of the ecosystem.

Then there was a need for housing both for the campesino families whose homes and land were in the areas to be reforested, and for the families of the workers who came in from other areas to help build the terraces and, along with the local residents, who also took part in the reforestation program.

Thus "Las Terrazas" came about, and by 1968, two years before the first Earth Day, was well established as what we would now call an ecovillage – a community very

Terracing the mountain sides helps stop soil erosion and is part of Las Terrezas' reforestation efforts.

consciously based on ecological design, restoration and economy. It has been working well enough to gain international attention; UNESCO declared it a Biosphere Reserve in 1985. The ecosystem is extraordinary, well worth visiting by anyone concerned about ecology or fascinated with nature. Here is one of the rarest of environments: a mountain tropical rainforest, and nearby, grassy savannas. Among its natural residents are 889 species of plants, some of them medicinal and many of them endemic, and 73 bird species including the tocororo, Cuba's national bird, and the world's tiniest hummingbird. The world's smallest frog lives here too.

Today we find a community of 850 residents in a picturesque village winding along the terraces bordering the San Juan River Lagoon. The villagers' primary work was, and is, reforestation. But recently, with economic hard times hitting the country, making the costly reforestation program difficult to carry out on the same massive scale as before, an ecotourism resort has been carefully introduced to provide additional employment and income for the village without negatively affecting the ecological priorities of the area.

The centerpiece of Las Terrazas and its resort is an ecolodge called "Moka" hidden among the trees and nurtured by the bubbling brooks of Sierra del Rosario. This mountain retreat was designed to blend harmoniously with its surroundings; it succeeds in many ways.

For example, everyone in the world of ecotourism talks about the objective of establishing positive relations between the ecoresort and the local community, involving the community in planning and assessment, and support

of the local economy, but it is almost never accomplished. In this, Las Terrazas is really unique – the interaction is total. The community was consulted and educated beforehand and approved the project. The community is already benefiting in many ways and gets 40% of Moka's profits plus 10% goes to health care in Las Terrazas.

The Las Terrazas Tourism Complex, which includes Moka, is now the second biggest employer in Las Terrazas; the forestry plan is still number one. All of Moka's workers live in Las Terrazas; only the administrator was recruited from outside since, obviously, no one in Las Terrazas had experience in running a hotel.

The hotel is providing additional food, medicine, and other supplies for the village. In addition, there is a restaurant with food provided by the Tourism Complex

The "ecolodge" called "Moka" is the centerpiece of the resort – the tree at the right passes through the patio and second story floors, out the ceiling and into the sky above.

for the residents of Las Terrazas. They buy a card that entitles them up to a certain amount – say 25 or 50 pesos – of meals a month at the restaurant – and they can invite their friends. Thus they become guests in the hotel along with the other guests.

A number of tourist-oriented home enterprises have sprung up – La Fondita de Mercedes, which serves typical Cuban food and cold drinks and Maria's place that serves coffee. There are craft workshops and art studios where artisans create silk screen, wood, ceramics, and recycled paper. Things that are sold in these workshops go 40% to the individual artists and 60% to the community. In addition, many visitors, inspired by this encouraging model of community development, give donations to the town, the school and the childcare center.

Most central to the educational aspect of Las Terrazas is its ecology center, equipped with meeting spaces for seminars and lectures. This center was established in 1974 and led to the UNESCO designation of Las Terrazas as a Biosphere Reserve. The Center's staff does exchanges with specialists from other biosphere reserves. Professional guides have designed interpretive pathways which branch out like spokes from the EcoCenter hub. The tourist trips are spectacular in their own right, with

beautiful walks along the forest/savanna edge, lunch from the ovens of the old plantation ruins, visits to orchid gardens, swimming pools and "massages" from the waterfalls that plunge into these ponds.

Best of all, everything Moka is providing is going to a community whose very existence came about through an ecological project of reforestation that is still going on today, and that will continue to the extent that the tourism complex is able to keep the community going. The terracing of the mountains and planting of native species that had been destroyed in the area not only brought back the trees, but by halting erosion, has also restored most of the original flora and fauna driven out by the last century's deforestation. Thus it is one of the rare ecotourism projects in the world where "eco" in the sense of both people and nature really does come first. Las Terrazas provides a model in which women and men are as important as the trees they plant and nurture, where each serves the other. As such, it is one from which people around the globe could learn.

Tom Whelan

Tourists enjoy swimming in jungle mineral springs and "massages" from waterfalls.

# EcoVillage at Ithaca –
# Creating an Environmental Model

## LIZ WALKER

"The ultimate goal of EcoVillage at Ithaca is nothing less than to redesign the human habitat. We are creating a model community of five hundred residents that will exemplify sustainable systems of living, systems that are not only practical in themselves, but replicable by others. The completed project will demonstrate the feasibility of a design that meets basic human needs such as shelter, food production, energy, social interaction, work and recreation while preserving natural ecosystems."

Those words appear in the Mission Statement of EcoVillage at Ithaca. To translate their intent into buildings and gardens, people and relationships, living economy and healthy ecology, five years ago we embarked on our journey. We began with a dream of creating a model of development that would address two of the largest problems facing us today: massive environmental devastation and social isolation. We chose to work on a scale which seemed manageable.

Anthropologists consider a village of 500 to be about the ideal size for maintaining diversity as well as good communication. The co-housing model pioneered in Denmark seemed to be an excellent social model, and we decided to organize a total of five co-housing neighborhoods clustered around a village green. We purchased 176 acres (71 hectares) of rolling fields and woods two miles from downtown Ithaca. Our environmental goals are to be carried out in our attention to land use, energy efficiency, and the integration of intensive organic agriculture. Although we are far from perfect, our integrated model is so widely appealing that we had hundreds of visitors a year even before anything was built on the site. We hope to inspire visitors, students and the general public with a new way of living that integrates the

best of traditional methods with the best of alternative technologies and social systems.

This means we are very self-consciously an educational organization. In addition, we have close ties to other educational institutions. EcoVillage at Ithaca (EVI) is fortunate to be affiliated with Cornell University through the auspices of our non-profit parent organization, the Center for Religion, Ethics and Social Policy. Although we receive no funding from Cornell, we attract professors and students who contribute in many ways. Some join in work days on the land. Others do senior projects or master's theses about EVI. Classes in anthropology, landscape architecture, city planning, ecology, hydrology, leadership training and others have studied our project. In addition, we usually have at least one student intern volunteering for us, often from other colleges around the country.

Co-organizing the Third International EcoCity Conference in Senegal was our biggest educational effort to date. One of our staff members, Joan Bokaer, spent a large part of two years working with colleagues in our sister village of Yoff to organize the conference. Much closer to home we also work with local elementary school

From across the new pond, Ecovillage at Ithaca housing under construction – roof shapes indicate south facing windows.

*Liz Walker is Co-director with Joan Bokaer of EcoVillage at Ithaca and was an early member of the planning committee that organized Ecocity 3. Here she speaks about the history and goals of her ecovillage, with a few comments added after the conference from conversations and correspondence.*

Coop member Jim Greuder lifts his daughter, Jamila, up the topsoil pile to play with brother Gabriel, Noam Sasner-Terman and Zoe Anderson.

Bill Webber

children through our Youth Garden Project during the summer and through hands-on workshops on the land.

We have an outreach program involving our members as speakers, hosts, writers, photographers and videographers. Part of our goal is to inspire others to take a fresh look at how to build a sustainable way of life and so, we host many visitors. Soon after Ecocity 3, for example, we brought Jeff Kenworthy to Ithaca to speak and urge people in our town to consider ecological ways of building. Our members regularly speak at local schools, clubs, and environmental organizations. One of us created a video about our project, and another created a slide show about the concepts of ecological city development. Perhaps most successful has been our media presence. We have been featured in many national newspapers and magazines, including the Washington Post, the Wall Street Journal, American Demographics, Popular Science, the Whole Earth Review and others. We see this as a way to get the message out to a wider public.

EcoVillage spent a year on land-use planning for its 176 acre site. Dozens of professionals and laypersons were involved in four Land Use Planning Forums which formulated our Guidelines for Development. Among the most important decisions was the one to preserve at least 80% of the site as open space: a combination of organic gardens, orchards, woods, wetlands and wildlife habitat. Two of our youngest members, Jen and John Bokaer-Smith who are here at Ecocity 3, have developed a 3 acre organic farm which operates as community supported agriculture (CSA) in which consumers pay for shares in

the harvest. This farm provides vegetables for 150 people. In addition we have community gardens and a Youth Garden.

In the winter of 1996-1997 our first neighborhood of 30 households will be approaching completion and residents will be moving in. We have adopted the cohousing model, which in our case too, emphasizes housing designed by the residents, homes clustered around a pedestrian street, and a Common House where people gather for evening meals, and social times. Over the next seven years we hope to create four additional neighborhoods. This highly participatory model takes patience and persistence, but it has a three-decade history of success in Denmark that resulted in very strong sense of community there. In America, several cohousing projects are successful after almost a decade and over 120 are either recently built, under construction or far enough along in the planning and execution to have site control.

At EcoVillage at Ithaca, homes in the first neighborhood are designed to be extremely energy efficient with passive solar features as well. Back-up space heating and hot water heating are carried out in energy centers which connect eight homes at a time. Many people will work on site, and a number of small businesses will be run out of the Common House. We have negotiated with the local bus company to provide a small shuttle van to downtown Ithaca.

Recently an EVI couple donated their share of the mortgage to make a conservation easement possible. This has transformed one of our dreams into a 55 acre

Bill Webber

Architect-builder Jerry Weisburd, wife Claudia and EVI Coop member Art Godin on site during construction.

(22 hectare) reality of enduring shagbark, hickory, paper birch, willow, cherry, aspen, hawthorn, highbush cranberry, honeysuckle, dogwood – and spectacular Finger Lake views. On this land, only preserved open sapce and farming will be legally permitted. From now on, the land can never be sold for other purposes. For the state's declining field birds, this is a tangible contribution and a model for future community development projects everywhere.

A last word should be said about local acceptance, problems and prospects. Though people concerned for the future are coming to Ithaca to live in the ecovillage from many distant places, and many local people participate in the CSA farm and are generally interested, the city of Ithaca itself is not especially supportive. As Richard Register has found in Berkeley, where parking requirements for new building result in ever more support for cars in spite of espoused environmental values and good recycling, Ithaca required that EcoVillage at Ithaca pave an unnecessarily wide highway-sized drive to our site from the main road. In many ways the city is decidedly not yet supportive of our effort. Tolerant – "if you do everything in ways we are used to" – would be a more accurate portrayal of their position, with little sense of the importance of building on ecological principles. If

we are interested in further advances and fewer compromises in our next four cohousing clusters, or if we are concerned for the establishment of more ecovillages in our area or for ecocity changes in the city of Ithaca, there will have to be changes in the way the government, and ultimately the people, regard city building in general, and our strategy will have to mature or change in some important ways.

*(Joan Bokaer has been working on this last challenge in the first months after the conference, as we will see under "Outcomes" in the last chapter of this book. There we will introduce her idea for more ecovillages and a streetcar line connecting them to new ecocity development and river restoration in downtown Ithaca.)*

Bill Webber

Tool shed at EcoVillage at Ithaca's organic Farm.

# The Earth is Our Habitat Campaign – Financial Leverage for the Pioneer Ecocommunities

## ROSS JACKSON

Gaia Trust's objective is to "promote a global consciousness that experiences the whole planet as a living organism and Humankind as an integral part of that whole." As to what that consciousness *does* as a result of that perspective, Gaia Trust has supported many active projects of organizations building ecovillages. Based on income from Gaiacorp, a company working in the field of currency management, and Gaia Technologies A/S, a commercial company investing in environmentally sustainable technologies and businesses, since 1990 the Gaia Trust group of individuals and linked organizations has been identifying and assisting ecovillage projects around the world. In turn, the principle people in these ecovillages have become part of Gaia Trust's active team of thinkers, builders, inventors, farmers and activists trying to transform human communities in balance with nature.

The Global Eco-village Network (GEN) evolved from 1991 - 1996 and was formally incorporated in Istanbul in June 1995. Here's the list of the "seed group" eco-villages from around the world: Findhorn, Scotland; The Farm, Tennessee; Lebensgarten, Germany; Crystal Waters, Australia; Ecoville St. Petersburg, Russia; Gyûrûfû, Hungary; The Ladakh Project, India; the Manitou Institute, Colorado; Auroville, India; IISF, India; Associacion Gaia, Argentina; Kibbutz Gezer, Israel; and the Danish Eco-Village Association. Yoff, Senegal may join soon. Gaia Trust has, as of summer 1996, spent more than $1,270,000 US supporting these living experimental communities in the conviction that there are real

Mudhouse in Ladakh, India, a member of the Global Eco-Village Network.

pioneers there, people putting their lives into communities and actually confronting the realities of a planetary ecological and spiritual crisis.

All this gets down to the specific technologies we adopt. Technology tends to determine the structure and organization of society. Contemporary society's alienating technology promotes unlivable mega-cities, separation of work and home, institutionalization of family support functions, environmental degradation, unsustainability, and overconsumption in a centralized, hierarchical structure. The GEN vision requires a radical change in structure that would reverse all of these tendencies. An important part of GEN's strategy is to promote sustainable technologies. The long-term vision is to provide sustainable jobs in ecovillages by technology exchange and cooperation. Three key criteria, in addition to commercial viability, have emerged in assessing appropriate technologies for ecovillages: 1.) ecological sustainability, 2.) human-scale, decentralized production, and 3.) allowance for a non-stressful, meditative lifestyle. (*Writes Dr. Jackson in one of the Gaia Villages publications, with a wry smile between the lines, "We realize that the realization of this vision will take some time."*)

---

*Ross Jackson did not attend Ecocity 3. But Gaia Trust, the organization for which Dr. Jackson is Chair, sent Thomas Mack and provided much of the funding that made Ecocity Builders' participation possible, including both important support for conference planning and participation at the event itself, and most of the financing for this conference report. Here we use his writings and others' from Gaia Trust brochures, "The Earth is our Habitat" proposal, Communities Magazine and other Gaia Trust papers.*

Ross Jackson, addressing conferees at the Habitat II Conference in Istanbul.

Ecovillages of the Global Eco-village Network tend to be primarily motivated by one of three concerns: ecological, spiritual or social. They all share the perspective that there is a global crisis in the way we build communities and share a commitment to changing that by building and living a positive alternative. Helena Norberg-Hodge, who works with Gaia Trust in a number of ways and heads up the Ladakh project in the Himalayan Mountain northern-most corner of India says we need models for the new millennium. A massive migration, the largest in world history, is going on right now – from the rural village into the city. It is not inevitable, she suggests, but rather "actively promoted and subsidised through economic globalisation and treaties such as GATT, NAFTA, and Maastricht."

"The idea that it is possible for the South to follow the Western model of urbanisation is highly unrealistic," says Norberg-Hodge (if profitable in the short term to those who promote it). With no savings based on past exploitation, *having been severely exploited*, "the countries of the South are driving their economies ever deeper into debt, exploiting resources at unsustainable levels, impoverishing their citizens, and abandoning their cultural heritage."

Nonetheless the raw economic prosperity of the North and its intensely persuasive commercial advertising and entertainment media make it a model. Therefore, we need a far better model emanating from the North: sustainable lifestyles based on the scale, technologies and values of ecologically healthy villages – and with it a self-critique based on what conscientious northerners have learned of their damage to living systems – things not mentioned in the advertisements and entertainment. People from the South can recognize familiar elements between their own ways and the ecovillages'. They can see that it is possible to be reasonably cautious and very selective about what should be accepted from the richer world and that some aspects, such as renewable energy and biodiverse technologies are helpful, healthy and fun. There's courage in the commitment to build and live in

ecovillages, too, risk changing one's own life and lifestyle, and this can also be persuasive.

Regarding that courage, the eco-villagers deserve the support of the international community because they are creating the examples that can eventually become mainstream and improve the quality of life for all of us. More than anything else, the world needs good examples of what it means to live in harmony with nature in a sustainable and spiritually satisfying way in a technologically advanced society. The best way, perhaps the only way to learn, is by example. The unique characteristic of eco-villagers is that they are out there doing it, with their own lives on the line. They are the pioneers of our time, breaking new ground, learning as they go, struggling against great resistance from mainstream society. And they are doing it with almost no help. On the contrary, our laws – zoning laws, tax laws, building codes, mortgage regulations – are often a great hindrance to attempts to live in a way that deviates from the modern industrial society model – a model that is unsustainable, but extremely resilient to change. Indeed, it may well be impossible to change it from the top down. This is why this eco-village movement may be very important for our common future.

A growing circle of ecovillage organizers including myself, my wife Hildur and Hamish Stewart have come together to provide just this kind of support, partially philanthropy from Gaia Trust, but mostly mutual self-help. The Global Eco-Village Network was created to share ideas resources and strategies, but Gaia Trust has decided that this approach is not enough. The crisis is too grave and the drift toward further destruction moving too quickly. We came up with another idea that we believe would accelerate the good works and make them much more easily accessible for others to adopt. It's a campaign we call "The Earth is Our Habitat."

Our proposal is that funds be allocated to an international United Nations committee for the support and development of from 50-70 small eco-villages of 50 to 2,000 inhabitants in both urban and rural areas across the globe. The primary objective is to establish successful examples of how humankind can live sustainably in the 21st century, with the intention of providing models for replication. The criteria for village-scale projects qualifying for the funding include: l.) priority for existing projects demonstrating past dedication, 2.) a spread of types of communities and diversity of locations around the world, 3.) a variety of different approaches such as the ecological, social and spiritual approaches, 4.) priority for projects exhibiting a wide range of appropriate technologies, and 5.) for projects with the potential to provide affordable, sustainable solutions to a substantial number of citizens in their region – the replicability factor.

The individual projects would receive between $0.5 million and $1.5 million and would be expected to use it on their projects in the course of four years. An evaluation would be due at that time, progress assessed, and new plans considered to implement what was learned.

Where the money would come from is not yet clear. Gaia trust has provided about one seventy-fifth that much money and seen considerable results, so in *principle*, if not at full scale, we know it works. Now we believe it is time to move much closer to the mainstream. Small is beautiful, said E. F. Schumacher, but small things can still be small yet big in number and benefit. That should be the case with ecovillages. If a consortium of foundations believed learning how to live in balance with nature were a high priority and village design, construction and technological outfitting were a means to that end, such funding could be possible. They could constitute a United Way for *building* a better future. (*This editor likes the idea of talking the World Bank and other major lending institutions into requiring that recipient nations and businesses using their funds must contribute substantially to the ecovillage fund administered by the UN Ecovillage Trust Fund (by whatever name) as a condition of receiving the loan.*) For

the moment, Gaia Trust is floating The Earth is Our Habitat idea as a powerful proposal for study, feedback and further refinement. Here's one place where a good idea – your good idea – could have enormous leverage. It's a challenge to all of us in the ecovillage and ecocity movement. Send your suggestions to the Gaia Trust, address in the resources directory.

An ecouraging recent development is that Wally N. Dow has given his endorsement to the Gaia Trust proposal and included it in his Habitat II report to the UN General Assembly in October, 1996. In addition, for its efforts exemplified by the Earth is our Habitat campaign, Gaia Trust was selected as one of the ten winners of the Habitat II World Business Forum's "Best Practices" award.

Courtesy of Gaia Trust

Domes at a Denmark ecovillage.

Fruit trees on the streets are rare and often opposed by governments, but here volunteers of all ages from Los Angeles Eco-Village in Los Angeles, California plant a small street orchard.

*As Jane Jacobs says, the city must fit into its hinterlands
(or bioregion) in a mutually harmonious manner. So too must
agriculture, sustaining humanity's community as it complements
and supports the diversity of life in nature.*

# 9. Sustainable Agriculture

There is nothing more basic for life than food. Eat and live. Don't and die. From microbe to human being. In the High-consumption World it is difficult to conceive of starvation, or that the sheer numbers of people can simply draw down the calories, vitamins and minerals faster than nature can put nutrients for plants – even with the help of industry – back into the soil. Or that pumping those chemicals into soils eventually degrades them seriously. For many of us, as we eat, the deserts spread, the seas become ghost towns. Children have a hard enough time comprehending that fruit comes from trees drinking in soil, rain and sunshine – not from peeling out cash at the supermarket. Children in the Low-consumption World have a difficult time comprehending that others can't understand that the origin of food is in soil, rain and sun, such an everyday labor it is to secure it, so much that obviously hangs in the balance.

Where this procurement of food and the recycling of its wastes into resources fit into the future city and village is the essence of our relationship with nature. Up until this century we simply went out and hunted species to extinction, usually for food. Now we extinguish species by even more means, including ever greater competition for habitats and soil. Poisons figure in, sprawl figures in. But so does transforming natural land to farms.

The question becomes, can we create a city that builds soil and enhances biodiversity? Our speakers seemed to think so. Joan Bokaer sees food production as the chief function of ecovillages, in healthy relation to their nearby cities. Thomas Mack, from New Mexico, describing permaculture, from Australia, portrays an agriculture redesigned, as the built habitat can be redesigned, to preserve biodiversity, build soils and restore forests. Amadou Diop describes Rodale Institute's Regenerative Agriculture Resource Center in Senegal. Coumba Bah, a young woman from Mali with a dream to deliver more and better food to the people of her country, describes what she believes to be appropriate food processing technologies for Africa. Lim Poh Im presents a grassroots effort in Malaysia to link organic farming with ecological land uses through the new western Pacific ecocity network she is working to establish. And Jen and John Bokaer-Smith, farmers from a Community Supported Agriculture farm at EcoVillage at Ithaca, New York explain how these institutions operate and why there are now 500 of them in the United States while ten years ago there was just one.

Together these talks and papers describe a sometimes new, sometimes revived relationship of agriculture to the city – and the village to the city. Let me propose that only by rebuilding the city, shrinking it back from most of the soil its sprawl has paved over, can we hope to make room enough for us all, food enough for us all, if that all includes the natural plants and animals. From now on we build soils as we build cities or our disasters will magnify greatly. It may be the ultimate expression of citizenship.

# Village Food Production for the City

## JOAN BOKAER

In nature, the by-product of one process always becomes the raw material of another process. Nature works in cycles, as Anders Nyquist described in his talk on Tuesday. And for us to sustain human life on this Earth we have to be able to design our habits so that all of our systems work in cycles. We need to learn from Nature. We will never achieve the supreme brilliance of the natural world but our goal should be to strive for maximum integration of all our systems so that there is no such thing as waste. Waste should not be in our consciousness.

This relates very strongly to agriculture at the edge of the city. In an intelligently designed city every human activity benefits the natural world. The simple act of eating and going to the bathroom can help grow food and if the agriculture is around the city, you then have this very important integrated relationship. When agriculture is far away from where people live, you lose the ability to integrate human activity in a positive way with the natural world.

So with that short introduction, I am going to show some slides. I will start with the Ithegau people of the northern Philippines who live in these beautiful steep hills and valleys with terraced fields working into and around belts and islands of forest lands. Their villages are nestled into an undulating tapestry of green. Their settlements are very compact because the vast majority of their land is for growing food. They have no waste. The three traditional villages here in this part of Senegal, Yoff, Ngor and Ouakam, who are hosting this conference, also

*Joan Bokaer was, with Serigne Mbaye Diene, co-convener of Ecocity 3. She is also the founder, with Liz Walker, co-director of EcoVillage at Ithaca just outside of Ithaca, New York, USA. In 1990 she conceived of and led The Global Walk for a Livable World across the United States from the Pacific to the Atlantic, from Los Angeles to New York City. She is currently working to get the city of Ithaca, New York to be the first city to adopt the International Ecological Rebuilding Program that was adopted at Ecocity 3.*

had no waste until recently and lived in a fully integrated way with their environment. Plastic, packaging and old car bodies are new here. Food scraps that the goats didn't eat were left on the beach for the tide and the fish – which the people would catch and eat later on. Waste is a new phenomenon here – which is why it's such a problem.

The city of Shanghai has 23 million people. At its edge, we see that compact neighborhoods stop abruptly at open fields. One thing that China has done well until recently – they are changing very quickly now – is that it has grown ninety percent of its food on the edge of its cities. Shanghai is a very good example, with their 23 million people they have a 50 mile green belt around the city. The farmers come into the cities in the evening, they pick up food scraps, they get human waste, they take it out to the country, they feed it to the pigs, the pig manure fertilizes the fish ponds, the people eat the fish and pigs, the fishponds provide irrigation for the field crops…. Everything is reciprocal. It's all integrated.

Now before the automobile, cities, like Tourette Sur Loup, the medieval town in the south of France in this photograph, were designed with an edge. They stopped and agriculture began. You can see the buildings clustered tightly, looking something like outcroppings of rock in a natural landscape. I spoke with someone who came from an Italian hill town, an architect named Tulio Inglese, who said where he grew up the people loved the land so much that it would never occur to them to put up a single detached dwelling. It just wasn't in their consciousness. And throughout history you see this concept of cities having an edge.

Let's look at a typical US city: Phoenix, Arizona as seen from an airplane. Barely a building is tall enough to be visible from this altitude but hundreds of thousands of acres of streets, yards, golf courses, parking lots – a cross-hatched landscape pattern literally stretches to the horizon. You find this kind of development on the fringe of every US city: sprawl. In this kind of development you lose your edge, you lose your wildlife, you lose the inner cities because it is so expensive to maintain them. You

Ray Bruman

An agricultural village in the Damt region of Yemen surrounded by fields of sorgum, millet, barley, corn, onions, carrots, melons, dates, olives, grapes, qat (Arabian "tea"), coffee and other crops.

lose everything. You end up with nothing. And this is how we have been building in the United States, particularly since World War Two, and much of the world is adopting our patterns and it is pretty frightening.

Now I'd like you to look at the plan that a developer had for the land that EcoVillage at Ithaca bought. On this map you can see the entire landscape is filled up with streets, driveways and an almost uniform pattern of very similar houses. We bought that 176 acres. It's a lot of land right at the edge of the city, just a mile and a half from downtown. This very large parcel of land was going to become 150 one acre lots – a typical American development.

Now look at our plan for EcoVillage at Ithaca. We proposed the same number of homes, 150, and the first of them are under construction right now. But we are putting all of them on 10 percent of the land. Look at the variety in this plan: clusters of houses in a few places on one part of the land, and agriculture, ponds, forest, and open fields everywhere else. Only one paved road and it goes to a common parking area.

What makes this an ecovillage? A big part of the ecovillage, for me, is that the community is producing food for the adjacent city. It is, in other words, redefining what is happening at the edge of the city. So the ecovillage in my mind is very much connected to a city.

In this picture we can see some of the fresh, delicious vegetables we started growing on our Community Supported Agriculture garden almost two years before we started construction. We've been using our "CSA" customer subscription system to deliver food to members, sell surplus to the public at a local food market and provide a place for people from the city to work in a serious, productive, rural garden. Low income people, college professors – all sorts of people dig in our soil. When you grow food at the edge of cities, what you get in the cities is especially fresh and nutritious – and there are many other benefits.

Now this is the city of Portland, Oregon, with its recently restored river-front park shown here. Portland citizens understood the importance of agriculture at the edge of the city. They saw what happened to Southern California and they decided to do something about it. They established an urban growth boundary around the city. And I highly recommend any of you here at this conference who are urban planners or ecocity activists to take this very seriously. In Portland they certainly do. There is no development outside this urban growth

Susan Felter

Children carrying fish from market. Yoff, Senegal.

revive a way to have agriculture work again as it once did, but also in the near future, in a healthy urban culture.

I want to finish with these fishing slides, images representing another kind of agriculture – fishing or aquaculture. I want to finish by coming back to Yoff for a glance and these stately tall women on the beach in their colorful print dresses called bou bous, with big bright red, blue, yellow, green bowls on their heads. Those big fish in the bowls with the crescent moon tails are the local tuna. These strong, gentle people still have a very powerful bond with the natural world. I strongly urge the government here to treasure, preserve and protect these fisheries. Don't let the European and Asian countries come in here with mechanical nets. There's a whole economy and culture here that comes from the ocean.

In our last picture we see a large crowd of villagers pulling the big fish nets in from the heavy surf. Here is the origin for our conference logo – the people working together, pulling in the net, bringing the food up the beach and into their village, to share with everyone.

boundary. They have created transit lines with light rail and all development occurs along the transit lines within the greenbelt. On the outside of the urban growth boundary is the agriculture.

I showed a plan of our EcoVillage; here's a photograph of one that exists. This one is high in the Andes and it has been there for one thousand years. Not everyone in this hamlet is a farmer. Probably only three or four people are. But look at these beautiful fields surrounding the village and imagine it is not too far from a city and busily supplying food for itself and the city. The idea of surrounding our cities with ecovillages is to

# Permaculture – an Ecological Design System

## THOMAS MACK

In the last forty years, we have lost 30 percent of the world's arable land, due to erosion by wind and water, fatigue from chemical application and salinization of the soil from faulty, misguided irrigation practices. Because we apply some 200 million tons of chemicals every year to get rid of bugs and produce our food, we suffer 200,000 deaths each year from poisoning. We employ this strategy, because we have a sense that we need to conquer nature. We label some things as good and others as bad, but without respecting all of life – including insects – we cannot achieve balance with nature.

As of 1995, we have lost 90 percent of the variety of cultivars that our grandparents once passed from generation to generation. In 1936, United States Department of Agriculture research indicated that Native Americans ate 1100 species of food. Today the average US diet includes about 25 foods. We must take diversity into our own hands by collecting seeds, propagating them and sharing them with each other. This is very important. The Irish potato famine is a prime example. Two million people died because they were dependent upon a single variety of potato. Today, we only cultivate a few varieties of potato in the US, but in Bolivia there are 12,500 varieties of potatoes. In 1995, we lost 27 percent of the US corn crop to blight, because we cultivate only three varieties of corn. In the Oaxaca valley of Mexico, where corn originated, there are 8,500 varieties of corn.

Our mono-crop mentality also has led to the decimation of many cultures around the world as we pursue and impose the Western industrial model. We are losing the diversity of culture, yet – to draw a parallel – we know that the diversity of the natural ecosystem is what accounts for its resilience and its strength. In the New Mexico area, the Native Americans understood the value

---

*Thomas Mack is an accredited permaculture instructor and professional permaculture consultant and designer working in Santa Fe, New Mexico. We have excerpted here from his talk at Yoff and from his later permaculture workshop report to the Gaia Trust in Denmark.*

of diversity. They planted corn, beans and squash together, calling the combination, "the three sisters." The corn provided the stalk for the runner beans; the bean provided the nitrogen for the soil; and the broad leaf of the squash provided the ground covering mulch. Planting things in combination can result in a synergistic system. Nothing exists in nature alone.

Australian ecologist, Dr. Bill Mollison, coined the word "permaculture" as a contraction of "permanent" and "agriculture" to imply sustainable agriculture as the basis for permanence in culture. Said Mollison, "Permaculture is the conscious design and maintenance of agriculturally productive ecosystems which have the diversity, stability and resilience of natural ecosystems. It is the harmonious integration of landscape and people providing their food, energy, shelter and other material and non-material needs in a sustainable way. Without permanent agriculture there is no possibility of a stable social order.

"Permaculture design is a system of assembling conceptual, material and strategic components in a pattern which functions to benefit life in all its forms. The philosophy behind permaculture is one of working with, rather than against, nature; of protracted and thoughtful observation rather than protracted and thoughtless action; of looking at systems in all their functions, rather than asking only one yield of them; and of allowing systems to demonstrate their own evolutions."

Permaculture design teaching and project work has established itself in over 70 countries throughout the world as a highly effective response to the planet's most pressing environmental problems. In that sense, permaculture has become a global grassroots environmental restoration movement, providing a positive, solutions-oriented approach to counter the destruction and degradation of the biosphere, Earth's biosphere.

As an integrated design system, permaculture draws on many earth science disciplines and applies "whole systems theory" to the functional design of ecologically sustainable living systems. The aim is to create habitat or living environments for humans, animals and plants that thrive on community and biodiversity, and function in synergistic relation with the earth's natural systems.

## Concepts for Permaculture Gardening

**Pattern & Edge** – using horizontal and vertical space to achieve the most efficient patterns for access, solar exposure, shade, condensation, water use and wind.

**Building Soil Fertility** – heavy mulching techniques for building layers of topsoil, nitrogen fixing species as intercrop and green manure cropping.

**Perennial and native species** – emphasis away from annual crops as the basis for meeting nutritional needs.

**Microclimates** – protected or specialized areas which can support plants/functions otherwise not viable.

**Guilds / companion planting** – planting for mutually beneficial species mix.

**Polycultures** – biodiversity generates dynamics and system immunity.

**Conversion of waste** – using cardboard, newspaper, tires, cans, etc. for garden functions.

**Optimal water use** – drylands strategies for water conservation and optimization.

**Nutrition based gardens** – food production planning incorporates local nutritional needs assessment and health maintenance planning.

**Food processing** – adding value at the farm; taught as part of the total nutrition cycle, from soil conditioning, fertilization, growing, preserving and preparing food.

**Forest Farming** – long term transition to predominantly analog forest systems to establish permanent agriculture and planetary preservation.

This holistic and ecological approach to the design and development of human settlement takes into account structures, technologies, energy, natural resources, landscape, animal systems, plant systems, social and economic structures. Permaculture draws upon traditional practices of earth stewardship from many cultures and periods in civilization, and integrates this understanding with appropriate modern technologies to serve the environment and humankind.

When considering the built environment, the design objective for applying permaculture principles is to create biologically responsive and ecologically sound systems. Buildings are designed with bioregionally appropriate architectural styles and vernacular motifs. They utilize organic shapes and spaces to promote health and well being. Structures and overall site layout are integrated into the surrounding landscape, so as to cause minimum destruction or disruption of existing biosystems. Building systems are evaluated with an environmental cost-accounting which considers non-toxicity, renewability, recyclability, embodied energy content (including manufacture and transportation) and other relevant factors. Anders Nyquist's presentation dealt very well with structures and their systems, so I will not delve further into those specifics here.

Permaculture can be a community-based construction process. A workshop brings together a group of people and creates an atmosphere in which participants can learn from one another as they assess the basic needs, cultural patterns and other considerations their project is to address. By designing and working together, communal functions are identified, planned and built in (with resulting cost efficiencies). The group members' assembled skills will determine which technologies are accessible and desirable. I will give some examples of projects and techniques which may be of interest to the people of Yoff and other arid regions.

In Santa Fe, New Mexico, we are considered to be in a desert region. (*Thomas begins showing slides of Santa Fe's piñon and juniper dotted landscape sweeping downslope from*

A permaculture/ecovillage imagined for a location in northern California, with compact living immediately next to nature and agriculture. There are workshops, offices, general store, solar greenhouses, organic farming and local, balanced production of fish, lumber and wine, for which the region is presently famous.

Richard Register

*the Sangre de Cristo Mountains into the vast grass, sand and cactus expanses of the Rio Grande Valley.*) We receive half of the rain that Yoff, Senegal, gets. Due to 300 years of over grazing sheep and logging, we have little soil remaining. This represents typical conditions permaculturists work with when using our design systems to restore a site by building soil, reserving water and reforesting.

Even in impoverished soil, we can grow grain as part of a system of continuous cover crops where the soil is never left bare. The cover of plants over soil is akin to skin covering the body – without the skin, the body cannot survive. The aim is to always keep the soil covered.

Permaculture emphasizes perennial crops which do not need to be cultivated each year. They return each year, and the system becomes self-sustaining. After six years of permaculture application in the desert, the soil supports a very diverse mix of species in five general categories: trees, shrubs, herbs, ground covers and root crops. The soil is built up until it is self-managing. There is abundance all around with food at every level within a permaculture orchard.

Waste water can be effectively treated using biological systems. I will quickly describe the functioning of a system designed to process the effluent created by about forty people in a desert setting and recycle it for farm use. The ponds are lined and filled with pumice, and filtering plants are put in the beds to take up the nitrogen, ammonia and phosphorous. The plants create biomes as well as clean water. Many of the plants have cultural value;

one of the plants employed was used by Native Americans for many purposes – food and basketry and so forth. This waste conversion system also enhances the value of the landscape to wildlife. At the site pictured here, just outside Santa Fe, we observed an increase in bird species from 29 to 65 species over just 3 years. We are providing a landscape and gardening system that provides food, not just for humans, but for animals, birds, bees and butterflies. We are also able to get beauty and economic value from cut flowers grown in this waste treatment system. A similar system could work well for Yoff. It would take care of human waste, create wildlife habitat, offer food for bees, provide compost material for making soil. The idea is multi-functionality.

In very severe conditions with little rain, one of the things we do is to capture water. Rather than let water evaporate or run off the landscape, we contour the land and create trenches which we fill with organic material. We then plant along that contour, and it then builds up the soil, allowing us to capture water which would otherwise run off. When erosion becomes critical and begins cutting away at the soil, we put in check dams to slow down the water, trap soil and organic debris and alter the micro-climate to allow for more planting.

I have great interest in a desert re-vegetation strategy developed by master gardener, Masanobu Fukuoka, from Japan who is now 85 years old. He is considered a national treasure in Japan. Over the past 50 years he has developed a strategy for re-seeding the deserts using a technique that we are now applying at one of the Indian

Courtesy of Birame Thiam

Permaculture workshop student
Birame Thiam with produce from the
Yoff farm parcel.

pueblos in a 600 year old village. We take clay, which is full of latent minerals, and we sift it. We combine it with some forest soil which contains forest floor fungi with the potential to form mycorrhizal associations. We mix the soil with seeds from a variety of trees, shrubs, flowers, nitrogen fixing plants, vegetables, etc. We then combine it with a bitter herb tea. We make seed balls with the mixture. Children love doing this, and they realize the potential of what is in these balls. The result is a package where the birds are kept from the seed by the clay, animals are repelled by the bitter herb, and the forest soil provides inoculant which spurs the seed to germinate. One of my flights of fantasy that could be more workable than it seems at first glance would be to use military aircraft, in this era of downsizing forces, to drop seed balls of appropriate design into desert fringe landscapes to reverse desertification and re-establish forests everywhere possible.

There is a great deal to cover in permaculture. There needs to be a great deal of exchange. I have worked in South Africa. There are permaculture projects in Uganda and Tanzania, and I would like to see this family grow.

(*After the conference, Thomas led a permaculture design workshop. The following is taken from his written report.*)

Our course was held from January 14 through January 25th, 1996 and was comprised of forty-five participants from the three villages of Yoff, Ngor and Ouakam. The course was conducted within the old village of Yoff at the headquarters of the Association pour la Promotion Economique, Culturelle et Social de Yoff (APECSY).

A few words of introduction are helpful here. The villages of Yoff, Ngor and Ouakam have been threatened culturally and physically by modern industrial developments. Factory fishing fleets have been so voracious that local fish counts have dropped dramatically. Predictions of the collapse of fishing traditions in Senegal range to within a few years.

Additionally, years of French colonialism brought the "green revolution" and cash crops to Senegal. This drastically altered traditional agriculture practices and devastated the generational information base about self sufficient natural farming. Huge monocrop plantations of peanuts depleted the land and deteriorated local nutrition, while primarily serving the French cosmetics industry. Aid organizations active in Senegal, such as the Peace Corps, are still recommending pesticides, herbicides and chemical fertilizers.

The population of Yoff has jumped from 25,000 ten years ago to nearly 40,000 in 1996. The village is running out of land to support new homes. Part of this problem is cultural, as polygamy is practiced and it is a sign of status to have a large family and more than one wife. Traditional village life incorporates large extended families in compounds. At the same time, one striking impression is that Yoff, with its strong cultural traditions, has much of the " social software" of an eco-village. The social organization and cohesiveness as a community is as impressive as anywhere on the planet.

In an environment like this, it was an honor and pleasure to be the lead Permaculture Instructor for our workshop, greatly aided by Assistant Instructor Moustapha Barry, a Senegalese student in the United

States who recently completed the Permaculture Design Course in Colorado. A number of Senegalese English teachers attended, some having been official translators for the Conference, and these participants were instrumental in translating all the course information into French, Wolof and local idioms.

The composition of the student body, with one third women, was highly diverse: gardeners, students, teachers, architects, engineers, mothers, translators, etc. Several youth associations from all three villages were represented, and quite a number of women's groups, including members of a women's dye and fabric making cooperative, a garden/farm cooperative and women's rural and community development groups.

After a few days of positive reviews within the village of Yoff, several elders showed up and continued throughout the remainder of the course. The elders offered contributions of tremendous value to the class, confirming the philosophy and describing how their fathers and grandfathers "did permaculture."

An excellent rapport was established with the Mosque elders. Many students in class had training in the Koran, in Arabic, and thus found passages in the Koran or among the Hadith (sayings and manners of the Prophet) which confirmed permaculture philosophy.

The classroom time featured two short lectures each day, covering the basics of the Permaculture Design Certification Course. Plenty of time was structured for hands-on activities and several observational exercises were conducted both within the village and throughout a 150 hectare property adjacent to the international airport at Yoff, which cannot be developed for housing and which the village is considering for an agricultural designation.

The village design exercise included inventories of unused resources and convertible wastes, housing layout, building design, water drainage patterns, traffic patterns and local vegetation. Field observation exercises on the farm parcel involved drainage patterns, erosion indicators, guild and biological associations, vegetative patterns, solar orientation and wind orientation.

The field exercises also included collecting, remarking on, and documenting local useful plant species. The class as a whole identified, categorized and notated 52 species. All the information from this was published in poster form for future reference, as were the notes from all the permaculture lectures.

Other hands-on activities included the construction of a "pee box" toilet for urine, which is simply a box stuffed with a sufficient amount of straw or other dry carbon material to absorb quantities of urine. The urine dries and rarely smells. The nutrient enriched straw is later used as mulch in the garden. There is a slow release of nitrogen along with the breakdown of the carbon material.

Many other activities were pursued over the ten day period including the planting of several stories of perennials, including mango, banana, papaya, and guava, with passion fruit climbing up a bamboo stake until the trees reached supportable size. We manufactured Fukuoka Seed Pellets for drylands regeneration and undertook design exercises for the planned Eco-Center which will serve as the model for the village extension area. This village extension will eventually serve upwards of 6,000 people.

By the time the course was winding up, people were already making compost at home, village permaculture groups were formed and the Khalifa stated that something "magical" was happening with this class – he was convinced that permaculture delivered the needed practical followup to the conference. Near the end of the course, the eldest participant said that all this reminded him of a story about the man who suffered an ulcer for fifty years, when along came his friend and inquired about the obviously painful condition. Indicating the stomach, the friend replied by asking the ailing man if he knew what the nature of that tree growing in front of his house was. Drinking tea from the leaves of the branches of the tree for a period of one month, the man was cured of his ulcer. "The answer is right in front of us," declared the elder.

# The Senegal Regenerative Agriculture Resource Center of Rodale Institute

## AMADOU DIOP

Since 1987, Rodale Institute's Senegal Regenerative Agriculture Resource Center has worked closely with farmers, researchers, and technicians to improve the quality of agricultural soils in Senegal. The long-term goal of the center is to increase food sufficiency for rural communities and decrease dependence on purchased inputs for food production. The project works to build institutional and community capacity to use and manage human and natural resources for people's well-being. The short-term goal is to develop partnerships with farm associations and government and non-government institutions. All of these work together in networking, education and training, and applied research activities to promote and advance regenerative agriculture.

We have four objectives serving these goals: l.) gather and disseminate information about regenerative natural resource management as it relates to food production and environmental enhancement, 2.) increase the capacity of local farmers' groups to conduct on-farm participatory design and evaluation of regenerative agriculture technologies, 3.) encourage farmers to communicate traditional and innovative technical information within the agricultural community to other farmers, technicians, researchers and, 4.) promote the multiplication of

*Dr. Amadou Diop is Director of the Rodale Institute's Senegal Regenerative Agriculture Resource Center in Dakar, Senegal. He began with a slide presentation and explained the work of his resource center. While showing images of daily agricultural activities, he proclaimed that "for the Senegalese in the countryside, the notion of waste doesn't exist. We live in harmony with recycling." In photographs he illustrated the integration of these activities into a community action and research/training project with examples of erosion control, cattle fattening, group meetings, manure processing, composting technologies and millet/cow pea intercropping. He showed women playing a key role in all phases of home, agriculture, and community life, planting, harvesting, transporting, selling, and cooking, often with a baby wrapped up snugly on their backs.*

regenerative agriculture technologies through educational media.

In rural areas where we have been involved, composting animal manure with crop residues has increased crop yield substantially. A major constraint to widespread adoption of this technique seems to be the unavailability of animal manure in adequate quantities. The center is encouraging farmers to integrate livestock with crop production in order to provide most farmers with sufficient nitrogen via organic material. During the last several years, reasonable millet and sorghum yields have been harvested from farmers' fields near cities. These fields were fertilized with urban waste collected by municipalities. This is an indication that farmers recognize the need to use organic fertilizers to regenerate degraded soils and increase their crop production.

The Senegal Rural Agricultural Resource Center has been working in collaboration with women farmers in home garden activities not only in rural areas but also in cities. Insufficient quantity of animal manure has also led farmers to look for other alternatives like the use of slaughterhouse waste as a substrate to grow vegetable seedlings.

Urban waste composting has become very popular in Senegal. The Center is implementing two programs in urban waste composting and home gardening in eight cities in Senegal. The purpose of these programs is to train people to make compost to be used in their own gardens or to establish small compost producing enterprises. The compost will be marketed to horticulturists or gardeners to satisfy their needs in organic matter around and within major cities.

The Center's gardening programs continue to play an important role in the improvement of family nutrition and income generation. Currently, programs are underway in Keur Banda, Gad Khaye, Koumpentoum, and Boundoum. The project seeks to increase family income, food quality and quantity, improve the nutritional status of these cities and neighboring villages, and reduce the use of synthetic chemicals in home gardening. Activities

Amadou Diop

Women participating in the Regenerative Agriculture Resource Center's agricultural training program.

include training in waste recycling, compost utilization, integrated pest management, and applied research.

A revolving loan fund has been established in Keur Banda since October, 1995. With the loans and revenues generated from their gardening activities, women are investing to improve the water supply in the village, to purchase goats for milk and compost production. In Thies, communities are being trained to manage a revolving fund program to give access – to women especially – to credit for implementation of regenerative practices. This program is a collaboration with a United Nations Development Program Project: Life-Local Initiative Facility for Environment. Through these and other projects, we believe Rodale's Regenerative Agricultural Resource Center is contributing to a reassessment and very healthy redirection of agriculture in Senegal.

# Appropriate Food Processing Technology

## COUMBA BAH

As we begin this brief exploration of food processing technology in Africa, I would like to first give a short definition of the concept of food technology. I believe that this is particularly important since many of you have been asking me in the past days if I am doing the same thing as Serigne. No, food science and technology is different from nutrition. We can think of nutrition as being more "health" oriented, and food science as "consumer driven." Food technology may also be "health" driven, but not necessarily. It's chief aim is at consumers' satisfaction and improvement of agricultural production.

In order for us to appreciate the importance of food technology in Africa, we need to glance at some aspects of our current condition. Those of us who come from Africa are familiar with these, but for the visitor it may be new information. First, production is seasonal – little longer-distance shipping means less variety – and we often see overproduction of perishable fruit and vegetable crops. Then there is negligence of traditional crops and methods of food conservation. Millet, among other traditional crops here in Senegal, is losing out to imported rice, for example; it is going out of style. There is a high degree of post-harvest losses and an inadequacy of food processing and marketing tools. Also, the markets are generally small and the people have low purchasing power. And, the cost is high for most imports that would otherwise be an appreciated and helpful supplement to local production. This becomes a base-line definition of poverty and it in turn causes malnutrition. Therefore, we can see that the importance of adequate food processing technology cannot be denied in our communities.

*Coumba Bah recently received her masters degree in food technology from Cornell University. While working in the Food Research Lab at the Geneva, New York campus, she focused on neglected African crops. When she returns to her native Mali, she intends to establish a food processing business. The writing here is a synopsis of her paper presented at Yoff, supplemented by details from a phone conversation.*

"But how are we going to get started?" some of you may ask. I believe that is indeed the right question.

The technology that we must develop and implement must first and foremost be appropriate; it should aim at maximizing the use of locally available resources. We need to transform perishable commodities into stable foods – which is exactly what food processing is supposed to do. We need to improve agricultural production, reduce post-harvest loss and increase "shelf life." Good food processing helps accomplish all these things and in addition increases the income and purchasing power of our people as it provides jobs and delivers a necessary product. In other words, it improves the economic system and creates services while meeting the demands it helps to increase.

In most of our African communities, small scale processing units are the appropriate solution. Generally they have low energy and low skill requirements. Specific technologies will fit particular locations, but not others, depending on local and regional resources, climate, soil, skills and available capital. Small scale processing units fit our present villages' scale, our rural agro-industrial operations. They add value to local commodities and fit the economic and social structure and the psychological and cultural behaviors. Among them are devices and containers for drying, canning, pickling, salting and making foods into sugared preserves. Solar energy dryers are obviously a natural in most of Africa. Home canning is an example of an accessible technology in many places and can usually be taught and organized with modest effort.

Now I will illustrate "the food pipeline" as it occurs in different parts of the world with an overhead projector image. In this chart we can see the diameter of the pipe is proportional to the amount of food flowing through the food delivery system, and the length is proportional to the level of processing. So in other words, the larger the diameter of the pipe, the bigger the amount of food processed and delivered to the hands of the consumer. The longer the pipe, the greater the value added.

Bill Mastin

The baobab tree is an important source of food, vitamins and wood.

The second case illustrated on the chart represents a developing country, where very little food is processed and very little value is added. The third case, is the worst scenario, the case of a failure. I have exaggerated this case from the point of view of the food processor that I am. In some places, other links of the pipeline are pointlessly overdeveloped relative to the capacity of the rest of the pipeline. To shift the metaphor, one strong link in a weak chain does not strengthen the chain. It is a wasted effort. This case, unfortunately, is what happened, and is still happening in most of our countries, where only one technical aspect of the whole food pipeline is studied and improved. Such cases must be avoided by all means.

The last case in our chart is the case of an integrated system. It is the optimal pipeline as I see it for African countries. The whole system is considered, studied and integrated. This is a system where one does not focus only on technical components.

Now in talking about implementation, let's look at some of the potential impacts that an appropriate food technology program could bring to a community when linked with good instructional programs such as training courses, workshops, market research and actual pilot projects, most effectively administered, I believe, through cooperatives. Let's take Yoff as an example.

Did you imagine or ask yourself what an adequately planned and organized food processing unit could bring to Yoff? I don't think that if such a program existed, we would have been so thirsty during our sessions here at the conference center. The women of Yoff already have the raw materials, the knowledge and most important the desire to produce superb beverages. What is lacking is adequate infrastructure to put all this together and make the juices available to us here at the center!

Growing up in Africa, I noticed long ago that imported and introduced foods were becoming more common as local and regional foods were disappearing. We have wonderful foods, among them mangos, papayas, millet and sorghum. There are some virtually unknown to the west. The baobab tree has highly nutritious and tasty leaves and its seeds are a tangy almost berry-flavored source of concentrated vitamin C. The bissap is a large plant in the hibiscus family from which teas and spices are made. The tamarind tree is truly extraordinary. Its fruits, leaves, bark and roots are all edible, it produces a dye used in textiles and its wood is good for building. But these and many other food plants are being replaced. There are several reasons. Due to long term colonialization, western influence has convinced many Africans that the imported and transplanted foods are superior and have better nutrition. In fact, there is no good information on the nutritional value of many African food plants because no one has bothered to study them. In addition, local crops are displaced because farmers are turning more and more to cash crops for export, such as peanuts and cotton.

Now, to close my presentation, I would like to leave you with a statement that comes from my own dear professor, Dr. Malcom C. Bourne. "Intermediate technology is directed towards helping people improve the quality of their lives by mastering and better using resources readily available to them, instead of relying on subsidized imported resources or resources that are out of step with their abilities to utilize effectively in improving their lot."

I do hope that we will all try to remember these words as we go about our future endeavors and struggle for a better world for all.

# Organic Farming and Ecological Land Use Planning in Malaysia

## LIM POH IM AND B. K. ONG

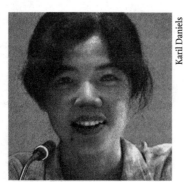

Lim Poh Im

Organic farming as a social movement is still in its infancy in Malaysia with only a handful of practitioners – most of whom are concentrated in a small community in Penang. The organic farming community was started in 1992 by the Penang Organic Farm, a non-profit NGO committed to promoting organic farming and ecological lifestyles. Apart from organising farmers to convert to sustainable farming, the group also runs campaigns and activities on a variety of environmental issues, one of which is ecological land use planning. Organic farming is promoted as a means to stabilise the serious urban drift which has seen half the workforce in the country switch to urban or industrial jobs in less than 25 years. During the same period, eighty percent of the fertile agricultural land has been converted to cash cropping under the influence of commercial and industrial interests. One of the consequences is the growing food deficit where food imports approached US$3 billion in 1994. The ecological "fallout" from such change is equally drastic in all its manifestations. All these concerns constitute important reasons for the organic farming movement to be interested in ecological land use – including urban design which caters to food production as well as the wider environmental concerns.

In order to find and stabilize a rural-urban balance, the Penang Organic Farm (POF) has started an information campaign on the urgent need for ecologically sound urban planning. These activities have included the presentation of a paper on the ecocity concept to an architecture conference at the University Sains Malaysia in 1994. A forum on ecocity concepts was organised for POF members following the presentation. Meanwhile POF has supported the application by a local heritage

conservation group for a project to revive an old mining town in the country by offering ecological planning input. In 1995, an inter-university architecture students' workshop adopted the ecocity as one of its major themes. A visit to the organic farmers' community was conducted along with a visit to a traditional house builder. Through these activities, the POF has developed a close relationship with lecturers and students as well as practicing architects and planners, heritage groups etc. POF currently is looking for opportunities to involve itself in ecological design projects to realise its vision. Our attendance at the Third International Ecocity Conference is aimed at networking with groups having similar interests and initiating an Asian Ecocity Network. It is also part of a program which involves internship training in an ecocity program administered by Urban Ecology, Australia, in the city of Adelaide as well as trying to integrate ecocity subjects into the university curriculum.

The POF sees the ecocity taking inspiration from traditional cultures as well as from the current ecological consciousness. The land use and housing construction patterns of traditional societies provide much convincing evidence of ecological considerations as well as popular participation. The traditional master-apprentice system can socialise building skills so that most houses are owner built units which suit the occupants, extended families in particular. Such a system should be adaptable to modern construction and planning, at least in part. If the traditional spirit of popular participation were brought to land use planning decisions, then society would develop a more democratic and participatory process and become more ecologically sensitive.

The pressing housing crisis in most Asian cities makes the ecocity a most desirable, if not an urgently needed perspective. Already, nature is hitting back at inappropriately-scaled and reckless development through a series of spectacular building failures, the Highland

*Based on a paper by B.K. Ong, coordinator, Panang Organic Farm, and a talk by Lim Poh Im representing the farm and the Western Pacific Ecocity Network. After the conference, Lim Poh Im went on to join the Halifax Project team in Adelaide and became an Urban Ecology Australia intern.*

Organic farmers and city folk relaxing in the first organic farmers' community in Malaysia.

Tower for example, and collapse of highways outside the country's capital such as the Genting Highland and Cameron Highland highways. Popular participation in planning should allow the harnessing of tremendous human resources to further ecological understanding which is most relevant in today's planning and design.

Apart from popular participation, many aspects of traditional rural lifestyles and values are ecologically sound. For example, people have been using composting toilets, mobilising the community in house construction (and also house moving), using local construction materials and techniques. More study of traditional living and research with traditional builders is needed before their valuable expertise is lost.

The Penang Organic Farm itself is situated on top of a hill in Balik Pulau. It is well worth a visit from any green tourists and all manner of students eager to learn about farming in Malaysia. The twelve-acre site situated 380 metres (1,300 feet) elevation is a pleasant 30 - 40 minute hike from the trailhead. It offers a serene atmosphere, cool, fresh and clean air, and a panoramic view of the entire West coast of Penang Island.

At the farm we grow herbs and spices, wonderful fruits, wild flowers and vegetables, and we attract birds, insects and animals which are part of the hill ecology. Organic farming does not use any chemicals whatsoever, and is based on ecological principles and methods such as mixed cropping, crop rotation and organic sprays. We provide free guided tours for public groups and schools.

Our farm was started by a group of enthusiastic young people who are committed to working for an ecologically and socially sustainable society. We leased a suitable 12-acre site where they could practice organic farming and joined the Willing Workers on Organic Farms – an international farm network which allows green travelers to link up with member organic farms located in many countries. Volunteers on the farm are welcome to stay at our traditional farm house which accommodates twelve people, or they can camp on the grounds. However, visitors have to bring their own sleeping bag and blanket. For a few hours of work a day, volunteers are entitled to free lodging and food.

Our present programs include public education in home-based organic farming and we publish a monthly

Courtesy of Penang Organic Farm

Cultivating food production as a revived career; a farmer
working at the Penang Organic Farm garden.

newsletter, supply speakers with slide shows and videos
to community groups and provide free consultation to
any group that wishes to start their own organic garden
plot.  A monthly organic dinner is held on the first
Saturday night of the month.  We have also formed a
Penang Organic Farming Club which runs a resource
centre with a library on organic farming and other wider
ecological and social issues.  The Green Store sells
literature, T-shirts, posters and videos as well as organic
produce from the farm and organic gardening ingredients
such as lime, fish and prawn meals.

# Community Supported Agriculture at EcoVillage at Ithaca

## JEN AND JOHN BOKAER-SMITH

For a city to be truly ecological, its residents must be in touch with their food supply. However, the inexorable growth of cities makes it harder and harder for urbanites to have any interaction with farms, farmers and the natural crop cycle. With urban growth resulting in ever-escalating land prices (and land taxes), small farmers are inexorably forced to sell to developers for housing, shops or office buildings, and food must be shipped ever-longer distances from ever-larger and more mechanized and chemical-dependent farms.

There are no simple solutions to the loss of cropland, and every ecocity should explore the full range of responses, including transfer of development rights, tax reductions for working farms, urban growth boundaries and other governmental mechanisms.

But beyond government action, there are also steps that consumers can take. One approach involves the development of community supported agriculture (CSA), a system whereby consumers collaborate with farmers by purchasing a share of a farm's expenses and, in return, receiving a share of the harvest. When a farmer has a whole community's support, instead of having one farmer, you have 100 families. In some cases the community comes together and buys the land to save the farm.

The greatest threat all farmers face is the unpredictability of the weather and the catastrophic consequence of drought and flood. By becoming partners in the farming operation, consumers share both the agricultural bounty and the financial risk with the farmers. In good years the shareholders enjoy getting more food for their money; when the harvest is bad, everyone shares the reduced production. Meanwhile, shareholders have the joy of knowing where, when and how their food is produced. Plus, their cost is less than if they shopped at a market for comparable items because there is no middle man and there is less wasted, unsold

*Jennifer and John Bokaer-Smith are farmers at the site of EcoVillage at Ithaca, in New York, who have organized and operate West Haven Farm.*

food. Finally, the produce goes immediately from the farm to the consumer without spending time under refrigeration or sitting in a warehouse.

CSA-style farming has been practiced in Europe and Japan for many years but is relatively new in the U.S. The first CSA farm in the U.S. was established about 10 years ago, but the number has since grown to over 500. CSAs are now considered one of the prime methods to save the family farm—and could become a key element in creating more ecologically healthy cities.

Our farm, West Haven Farm, is part of EcoVillage at Ithaca in upstate New York. We are members of the group that owns the village, and we lease our three acres of farmland back from the full group. We have been in operation for five years and fortunately, at the moment, the lease is at a nominal cost. Here's how it works.

There are presently 35 full shares in the CSA; some families double-up on a share, meaning that between 50 and 60 families actually participate. A share costs $410 a year, for which a family gets a weekly pickup of produce for about 25 weeks. The cost is roughly wholesale – each shareholder gets about $600 to $700 worth of produce over the season. Each week, we harvest what we figure will be a usable quantity, and the family gets a choice of about seven to ten different items, depending on what's ripe.

We survey people to find out what they like. Over the season we grow and sell about 40 different crops. The few foodstuffs we don't grow well or don't have the space for – corn, squash and strawberries, for instance – we contract out, and pay other farmers to grow. We try to meet people's full produce needs for that week. Our goal is that our shareholders will not have to buy any vegetables that week. For the physical distribution we've used the garage of our old house downtown. Now that we live on the farm itself, we plan to arrange two different pickup sites—one at the farm and one downtown on someone's porch or in their garage.

The original concept was that the food was individually boxed up and that every shareholder family would

Bill Webber

Jennifer Bokaer-Smith, one of the two farmers who organized the West Haven CSA Farm in Ithaca, New York.

come and get their box. Now, to make it easier on us, every Tuesday we put out big displays of vegetables and fruit with instructions as to how much of each to take. People bring and reuse their own boxes. We have a chalkboard with instructions and a checkoff system. If a family is out of town they sometimes get a friend or neighbor to pick up for them, and often these new people get so excited about the concept they join themselves.

To be fair, community supported agriculture isn't for everyone. CSA shareholders don't necessarily get exactly what they would get if they went to the market. People who cook only directly from a recipe are usually frustrated by the lack of choice that they may be used to; it works better for people who like to open the refrigerator and say, "OK, we have a lot of turnips, let's cook something with turnips!" There is a bit of waste because someone might get something he or she doesn't like or doesn't know what to do with. To deal with this problem we lump things together—root crops, for instance, or greens—and say, "Take 10 pounds of roots" or "10 pounds of greens." Then they can take what they like from among potatoes, carrots, beets, squash, onions, or from among spinach, kale, mustard greens. It's halfway between the total choice of the supermarket and complete lack of choice.

At West Haven Farm we've done very little marketing of the concept. Actually, in the first year it was more like a big garden. Both of us had regular jobs, so our farming was early in the morning and late at night. We had both apprenticed on some farms and knew something about agriculture, but we couldn't commit ourselves financially. So we made a list of friends who were interested in fresh food. Depending on how much we picked each week, we would call that far down the list, invite them to come with bags, they'd come and pay us $10. Next year we had more land, got advance commitments and sold shares. It just grew from there by word of mouth.

Some CSAs jump right into it and market heavily. Or they combine with existing community or church groups to get started. Once established, CSA farmers can spend up to 90 percent of their time farming, while conventional farmers find themselves spending at least 50 percent of their time marketing. Even though we also have to do some marketing, the difference is that most of ours takes place in the off-season. We get our subscriptions for the coming year during the winter months when there's not that much to do except make repairs. When the season starts, we don't have to do the marketing.

Some of the best things about CSAs do involve some hidden costs. One is educating consumers about how to cook vegetables they aren't familiar with. We give all shareholders a cookbook and also have a weekly newsletter with recipes and news from the farm. Also, people have to learn to eat with the season. They have to learn about the agricultural cycles. A CSA brochure usually has a big graph with the months of the year and an indication of the crops that are harvested each month.

We have no work requirement. However, we do offer people the option of working for part or all their share fee. Right now about seven people work off anywhere from $100 to their full share in a season. Generally they come once a week from something like 6:30 a.m. to noon. It works out to $4 or $5 an hour—not much but it's better than we're making ourselves! We also organize volunteer work days where people can come and help with the planting and harvesting. These are social events that help us in time of big need. As we always say, people aren't just buying the food, they are also saving the farm.

Community supported agriculture brings the farmer stability. We had a drought last year. It was devastating, but the CSA saw us through. The shareholders knew that there was a risk and that they would have some lean years. Fortunately, this year was bountiful and the shareholders did very well. In New York State, the only farmers who are making a living from farming are those who have inherited a farm or all the equipment. Farms today can support themselves if they are already set up, but they can't cover the cost of the land or the infrastructure. We've been able to build up, because we have other jobs. In fact, nationally 90 percent of farm families get over 50 percent of their income from non-farming activities.

Virtually all CSAs are organic farms. It's only natural: when the farmer knows who's getting the food and when the consumer knows where the food is coming from, then organic is the obvious way to go. We feel good about growing organic food for our shareholders. Conventional agriculture has many hidden costs, such as pesticide residues in food, poisoning of farmers and farmworkers, soil erosion and nitrates and phosphates leaching into groundwater and streams. As an organic farm, we rely on cover crops and compost for healthy soils which in turn grow healthy nutritious plants. We count on cultural, biological and botanical controls to moderate insect and disease problems. We operate at a scale where we can do most of our work by hand, contrary to common mechanized farming which removes jobs from the farm.

There are many different models of CSA farms. We're on one extreme of the spectrum where we do everything—farming, marketing and distribution. The other extreme is when a non-farming group of city people come together in search of good organic food. They then collectively find a farmer or even hire a farmer to grow for them. Usually it's somewhere in the middle: there's a farmer and a core group of organizers and they all work together. We're actually now trying to start up a core group to share our concept with. We've found that we can do the work but that we need more advocates—we're not very good advocates for ourselves. Part of the concept of CSA is that farmers should be able to make a living wage and we're not there yet. We've always priced our share on what people would pay rather than what things really cost. It would be easier to price the food more realistically if we involve a larger group of people. We want to ask the core group, "How can we meet your needs and also make this farm work financially?"

We also want the help of a larger group in envisioning where the farm should be going. Should we extend the length of our season by constructing greenhouses? Should we involve church groups?

Our goal is to support ourselves and our farming habit. We would be happy if our shareholders were our sole outlet. In fact, however, about one-third to one-half of what we grow – depending on the weather – is sold at farmer's markets to the general public. The reason we've had to find other outlets is that the community supported side of our operation hasn't grown fast enough.

We see ourselves as part of a growing movement of young people who are interested in farming. It's a great job. It's very hard work but we get to make decisions each day about what we will do. We get to be outside all day, which contributes to a very high quality of life and high job satisfaction. Even though we may not earn as much money as most of our friends, we are happier in our work.

Oxen-driven water wheel at Balat village, al Dajla Oasis, Egypt.

*Consciously or not, the community has a self-imposed
limitation on the quantity of water available, thus avoiding
exhaustion of the aquifer which feeds the well.*

Mariano Vazquez

# 10. Ecotechnology – Tools for Harmonious Living

The idea of "green" technology as a galaxy of associated industries, businesses and lifestyles, "environmentally responsible mutual funds," "simple living" approaches and so on, has been much debated. The technologies themselves, from traditional organic agriculture, say, to solar photovoltaics, are fairly well developed in various places around the world. Architect William McDonough suggests that we should be launching a "Second Industrial Revolution" based on nature's strict rules of rigorous recycling. Technologies that turn wind to energy and garbage to rich soil, ones that use low "embodied energy" and materials of local origin for textiles, furniture and above all buildings, take us a long way toward protecting the biosphere and sharing benefits with future generations – the bottom line for "sustainability."

But the debate continues. Are electric cars "green" if they perpetuate sprawled land use development, continued highway carnage and demand massive new energy supplies? Are taller buildings "green" if they are solar heated and lighted, cloaked in greenhouses and designed with terraces like the emerald mountain-sized agricultural stair cases of Bali and the steeper, drier ones of Yemen?

We can't know if many technologies really are healthy until we view them in relation to what cities and villages can become when and if we understand their nature and potential in depth. Conversely, without exploring the technologies themselves, we can't know how to design and build the villages and cities.

In this chapter we view technology from several different angles, and if not always strictly in relation to the ecocity, to the greater education of those of us who would be ecocity builders. For such are the surprises of a good conference, in which, as you will remember, the organizers don't get to write the script. Mariano Vazquez explores Egyptian and Spanish traditional technological choices from the social perspective: lots of people prefer views over energy conservation, alas, perhaps – or perhaps not alas. Donna Norris builds the edge of adventure right into her Northern California solar houses, mixing warm, civilized sunlight with the dangers of nature – on cliffs above the churning ocean or in sun-facing forest houses with cliffs built in. Elizabeth Leigh takes us to a low-income neighborhood of Washington, D.C. to look at the design process itself, as inner city high school students explore design in ways designed to open their eyes to both their own and their communities' potential. Samba Lô gets down to the gritty particulars of solving part of the solid waste problem of his native Senegal and some of the energy problem at the same time: by making burnable briquettes from garbage. Harriet Hill from the United States Environmental Protection Agency refers to a whole world of experience in turning liquid sewage into not just soil-building fertilizer, but habitats for enormous numbers of species, both endangered and smugly successful. Architect Kirsten Kolstad links past with future, tradition with energy efficient public-assisted housing. There, in Oslo, Norway, refurbishing buildings becomes a substantial model for a healthy future.

# The Role of the Inhabitants in Ecological Approaches to Architecture

## MARIANO VAZQUEZ

I want to begin by dedicating this presentation to the memory of the Nigerian writer Ken Saro-Wiwa, killed for defending the culture of Ogoni people against the interest of transnational companies.

My aim here is to recall the role people must play in any ecologically based project if it is to be successful.

Two hundred years ago most people lived and worked in buildings they made themselves. As we know, industrial society changed all that, producing houses through industrial processes, concentrating people in large populations, intensifying sanitation problems and so on. In this process, industrial society's attempt to provide lodging for the masses ended up affecting both rich and poor, and housing became a technical problem.

The Modern movement was probably the most complete attempt to solve the housing problem. It substituted the physical and cultural reality of the international style – from design and use of products to architecture, landscapes, city layout and transportation systems – for whole vernacular cultures. This happened wherever the international style was adopted, which was almost everywhere. In this approach, the "house with exact respiration" was proposed for all countries and climates. So called "neutralizing walls" – two sheets of glass between which the artificially conditioned air would circulate – would maintain a constant temperature of 18 degrees Celsius (64 degrees Farenheit) inside the building. This demanded a hermetically sealed house, a house incapable of relating climatically with the outside world, a house "against" a dominated nature.

In the Domino House, Le Corbusier broke up the structural functions of the building, separating support structure from the many other functions of the wall. He could do this because reinforced concrete column and

*Mariano Vazquez is Associate Professor at the Polytechnic University of Madrid and co-director of a postgraduate course called "City, Architecture and Ecology" in the School of Architecture. We have synopsized from his slide presentation and his accompanying paper.*

slab construction was strong enough to span large areas with no walls at all. Walls made of more traditional materials provide not only structural support with a continuous wall, but also serve to provide shelter from wind and precipitation, insulation, moderation and retention of heat, control of light with well-placed windows and doors, provision of useful inside wall surface and so on. Le Corbusier was the first to develop an explicit theory differentiating between the structure and the wall of the building, extracting the structural function of the wall from the other functions.

There was no valid historical precedence for this type of dissected construction in which it seemed sensible to have a specific approach for each particular problem, one solution for spanning space, one for insulation, another for lighting and so on. It amounted to an approach totally separate from the whole-systems process of perception and construction which was the basis of the vernacular cultures that were being replaced.

I want to underline the very destructive influence of this process and the International Style of architecture that employs and promotes it.

In the megalopolis of the so-called Third World, very little housing is provided for the poor, and in some places, none at all. The poor of these urban areas must recourse to self-building in order to obtain shelter, but they cannot fall back on the vernacular traditions because the community that could have maintained them has since disappeared. As a result, the International Style ideal concrete and block or brick dwellings that do get built are virtually identical everywhere. Photos of these buildings taken in different cities of separate countries and continents offer no clue to their location. Examining some problems and their solutions can help us to see clearly the line between the technical solutions and the suitable ones.

### Ecological Technicians – Al-Qurna

Hassan Fathy is well known for his passionate attempts to reconstruct peasant life and community architecture in Egypt. His most important project in this

Mariano Vazquez

The mosque was the only building maintained at the new town of Qurna, Egypt.

regard was the new town of Qurna. The old city was constituted of five villages located in the city of Thebes of antiquity. Early in this century, tomb robbers lived there and prospered. In 1944, an enormous catalogued item, a real archeological treasure, was stolen from the ancient necropolis. The Ministry of Antiquities retaliated with a decree for expulsion of the inhabitants and expropriation of Qurna itself.

Fathy was given the project of building a new town and delivering new housing as indemnity. This was a new opportunity for Fathy to demonstrate the feasibility of his reforming ideas. The construction would be collective. The participation of the people would be included in the design process and in the recovery of old building techniques which had fallen out of use, such as the construction of vaults and domes of adobe. The construction would extract both practical possibilities and artistic order from ancient tradition. Fathy wrote about these techniques, which are still in use in Upper Egypt, transmitted by builders from generation to generation. But despite the ability of this tradition of construction to

solve many social and economic problems, it has been disappearing due to lack of interest – a situation Fathy hoped to remedy.

The construction of the new town was plagued with administrative problems and incidents involving the future residents. After many problems, by the mid 1960s, the unfinished village was eventually inhabited by people from all kinds of origin, mainly squatters. By 1982 the new settlement already showed notable symptoms of growing old, due mainly to the lack of maintenance of the buildings which, in my opinion, showed the absence of care and involvement of the inhabitants. The only building that was maintained was the mosque.

Ten years later, the settlement had suffered several alterations. The theater had been restored under Fathy's supervision, but it was not used by inhabitants. The primary school had been replaced by a new concrete structure. The houses had aged badly or had been replaced – by precisely the humble version of the International Style which we have seen before: concrete frames and brickwork or concrete block.

In this project which aspired to be exemplary, Fathy failed to achieve his most intimate desires, as he ably expressed in his book, Construire avec le Peuple (in English, Construction with the People). It is not difficult to see where the error lay. As Fathy himself indicated in the book, his methods consisted of integrating the vernacular techniques within the catalogue of solutions of his own architecture. But it remained the technicians who presented the problem and studied the solution. The archeologists solved the problem of the Theses of the pharaohs by expelling its inhabitants; Fathy designed with his best will and intelligence what he thought would be a new town adapted for the people. But the former inhabitants of the old Qurna suffered from foreign solutions, and in fact, did not have sufficient opportunity to decide their own solutions, despite Fathy's original intention that they should. Actually, it is difficult to see how the vernacular techniques can be useful when the communities that gave them life have already disappeared.

### Ecological Technicians – Balat

The Saharan oases, which appear in topographical depressions, will be the next example. Along their cliffs there are many opportunities for the subterranean aquifers to emerge, creating natural springs which can irrigate large extensions of land. Furthermore, it is easy to reach the aquifers by means of artificial wells.

Balat is a small village in the oasis al Dajla. Something strongly impressed me when I visited in 1983, but I could not put my finger on it when I was there. After I returned to Spain, while looking over photographs of the trip, I realized that Balat had the only clean (by European standards) coffee shop that I had seen in Egypt, though I had enjoyed many. The photographs I contemplated clearly showed the loving care of the building, the placidity and freshness of the gardens in the hot hours of the middle of the day. It was all there in a photograph of an empty coffee shop in the oppressive heat of a summer afternoon. In Balat, as contrasted with places along the

A clean, well-lighted coffee house in the oasis village of Balat, Egypt.

banks of the Nile, the people were, to the extent of their powers, their own masters.

I will try to show this difference by describing the use of water. The gardens of Balat are irrigated by artificial wells. The water is extracted from them by water wheels driven by oxen which are located in small buildings that protect the animals and the people in charge from the sun. The system for lifting the water is a series of ceramic vases tied to a double cord and limited to a certain depth of water extraction by the friction of the system and the strength of the oxen. That limitation has a significant effect. When the water level drops below the maximum depth of the particular technology, irrigation of gardens has to be abandoned. Consciously or not, the community has a self-imposed limitation on the quantity of water available, thus avoiding exhaustion of the aquifer which feeds the well.

Alongside these water wheels, new pumps have been installed using development aid funds. But in Balat, they have not been used, nor were buildings erected to protect them from the weather. Their use would have allowed overcoming the limitation imposed by the water wheel system and would have significantly increased the amount of water available. But, aside from whatever were the true reasons, the fact is that whilst things remain as they are, the community of irrigators of Balat is unlikely to face the complete collapse of their aquifer, and hence their way of life, due to the kind of over-extraction that is typical in the industrialized and rapidly developing world.

The photograph I am showing you now is from the Oasis of Albahriya, one of the many oases in which "modernization" projects have been carried out, projects of the sort that Balat simply ignored. Here we see two taps emerging from a sand dune. I never found out the reasons for such a mysterious installation. Whatever it was, I think the image illustrates well the perplexity that these "modernizing" projects can occasion, and the impossibility of truly understanding them.

### Ecological Technicians – Patones and Pelegrina

In our day, many of the technicians interested in solving our ecological problems have grouped their efforts under the heading "bioclimatic architecture." And within this tendency, the solar orientation of buildings is considered to be of crucial importance.

Without a solar orientation appropriate to the seasons, the building or the town under analysis will not be considered "bioclimatic." But, is it really a crucial question? It is certainly easy to see that it is not thus in general. Certainly in Spain, there are cases of old villages in which it is not reasonable to doubt that the vernacular builder was influenced by solar orientation, such as at Patones. The plan of this village appears adapted to the winter solar course, so that none of its buildings remains in the "zone of shade" projected by the surrounding mountains.

But if one wants to prove the contrary, that is to say that the vernacular construction has not always considered facing the sun as necessary, one can also find examples. Pelegrina is such a case. The River Dulce circulates through a rocky defile to the south of the village, but opens to a beautiful valley to the north. Buildings in the north-facing quarter do not benefit from the sun in winter that graces the south-facing quarter, but they have views of the valley.

Like many other villages in the interior of the Iberian Peninsula, Pelegrina has suffered a drastic population reduction, though recently some people have returned or new people have come. In this process of minor recovery, both quarters have benefited from the preferences of the public. Amongst the newly restored buildings the only house with a typical "bioclimatic" device, a solar collector for hot water, has appeared "where it should not be," in the cold quarter. This structure has a shocking roof in which the collector faces south whilst the roof faces north. The heating is provided mostly with evergreen oak logs from the communal mountain during at least eight months of the year. There is only a single small window in the north wall which frames a fine view of a waterfall on the other side of the valley where vultures glide. Can we consider this house an "ecological" dwelling? If the answer is affirmative, the reason cannot be a technical one and we will need to search for it in the autonomy of decisions of the builders.

### The Limits of Technics and the Role of the People

In the above examples I believe I have illustrated the limits of technics in designing "ecological" practices and artifacts. In this conference we find people from very different places and cultures, trying to learn how to rebuild our cities. I think that the attempt to solve this problem technically clashes continuously with an insurmountable barrier. I also believe that recognizing this barrier is essential for the success of our aims. Therefore, I will try to conclude by describing of what this limit consists.

*(At this point in his presentation Marino critiqued the method of decision making typical in technical evaluations that seek optimization of key factors, such as financial savings, energy conservation and so on. But he pointed out that further evolution of the very theory of optimization used by industrialists and many planners and managers have shown that the majority of our "real" problems are unsolvable ones technically. For example, in mining and many other industries, optimization of expenditures in a particular process (saving money) often is contrary to optimization of worker safety (saving lives). The more money is saved generally, the less lives, and vice versa. A technical decision is impossible; a values-based decision from human experience is called for.*

*(However, he says, the technicians can have a role in the non-technical process of decision: to eliminate those solutions that are decidedly worse. But it must be the people who are involved in the process who select the appropriate procedure.)*

If the people do not lead the process of ecological transformation of our cultures, then there will be no socially accepted, hence applied, definition of what is "ecological." Leadership for ecological design and building is much more than the easy recourse to the participation of people. It is above all the decision itself which is the source of power, information, and control about one's own future and that of the surroundings. It is the decision that makes the difference, as Fathy says. We cannot keep on ignoring that the ecology of human societies is the evolution of their consciences.

Mariano Vazquez

Unused water faucets at the Oasis of Albahriya – the mysteries of modernization.

# Life by the Light of the Sun – One Solar Designer's Approach

## DONNA NORRIS

"Marvelous in a room is the light that comes through the windows of that room and that belongs to the room. The sun does not know how wonderful it is until after a room is made." – Louis Kahn

I know how it feels to have the sun in one's home. To have it every day, in ever changing ways, in every corner of one's existence. To have it sensitize one's ability to tell the time of day and the seasons of the year. To have its energy controlled by thoughtful solar design so that its presence contributes to natural heating, cooling and lighting. For I am blessed with two solar houses.

One of these, my parents' home, stretches out in a sculpted coastal bluff. The other, the one I share with my husband, reaches up from a ridge among the remains of an ancient redwood forest. Except that both are long and thin, at first glance these houses bear little resemblance to one another. Unless told, the uninitiated would probably not even recognize them as solar designs.

But once inside it is obvious these houses have much in common. They are quiet, warm, light and bright. One observes that the house plants seem to be particularly happy. Even the most ordinary varieties blossom. Our pets would not think of "running away from home" and people who visit do not want to leave. My parents, both in their eighties, credit life by the light of the sun with their increasingly good health. There is a sense of general well-being "in the air."

Curious to know why these houses feel so good, two years ago I began studying their patterns of warmth and brightness, the configurations of their spaces. Knowing that these are extreme houses, intuited and largely experimental, I coupled my observations in them with research and visits to other solar homes. My hope was to discover ways that our daily gifts from the sun could be

Donna Norris

Donna Norris' residence, her tower house, with qualities of mystery, refuge and danger.

made available to all through a renewed approach to architectural design. As energy savings have not been a compelling enough reason for people to seek solar homes perhaps their beauty and healthfulness would be?

I have not been disappointed. My findings indicate that solar oriented spaces are not only obtainable but also infinitely desirable. Apartments advertised as "sunny and bright" are always the first to be rented and generally for more money. Subdivision houses that "by chance" have a solar orientation are always the first to be sold simply because they have more appeal.

But whatever their form or wherever you find them, successful solar space designs are based upon a set of patterns. The patterns include but are not limited to: the sun orientation, controlled entry of solar energy, south facing windows, structural thermal mass, insulating envelope and connection with the Earth.

---

*Donna Norris designed and built two solar homes near Mendocino, California that she discusses here; she is currently planning a solar neighborhood of 23 to 28 houses in Jacksonville, Oregon, USA.*

Bill Mastin

"Heavenly light" suffused Shaker interiors
at Hancock village and in the Shaker
room pictured, at Pleasant Hill, Kentucky.

The more of these solar patterns one is able to incorporate into the design and construction of a space, the greater the benefit and joy to its inhabitants. But to be most successful, how each pattern is incorporated must be based on thoughtful design considerations including an analysis of the climatic conditions and geographical location of the building site. In our homes we are surrounded by warmth and natural light. We also save money. But these things alone do not explain their appeal. When the form of a house grows from these solar patterns, delights that are rarely on anyone's "wish list" when they are designing a house, appear as if by magic.

One morning while reading Grant Hildegrand's, *The Wright Space*, I became intrigued by the possibility that the characteristics he describes as being "innately preferred by people in both their manmade and natural environments" might occur naturally in solar designed houses. He listed these characteristics: Prospect – a place from which one can see a considerable distance. Refuge – a place that provides shelter or a place to hide. Order – categories of regular, harmonious arrangement of elements. Complexity – delights and differences within each one. Hazard – a source of danger or assumed danger. Mystery – a secretive quality.

I have found numerable examples of these characteristics in my houses. Prospect and refuge occur in both interior and exterior forms. My favorite examples of order and complexity are those created by shadows and reflection; hazard and mystery by the many hand-crafted stair and ladder ways. Admittedly these may be unique to my solar house and, I might add, not recommended for the faint hearted or acrophobic.

Architect Christopher Alexander describes the buildings he admires as those which possess a "quality without a name," buildings which are beautiful, ordered and harmonious, buildings which live. Surely you have

seen examples of these? A small cottage with an old garden filled with sunlight and vintage roses. Lighthouses on their stony ramparts with bursting waves.

Alexander maintains that it is possible to design new buildings which have the "quality without a name" by including what he calls "living patterns" or architectural elements. More than 150 of these are described in his book, *A Pattern Language.*

When I looked at my solar homes for examples of his patterns I discovered many. They seem to have grown out of the solar patterns around which these houses were built and they include: long or tall thin house, indoor sunlight, natural light on two sides of every room, thick walls, radiant heat, tapestry of light and dark.

In a further search for answers, I recently read through my personal diary and noted an increasing number of references to nature since I began living in these houses. It seemed reasonable that part of their appeal might be explained by my increased connectedness with the earth, sky, sun and stars.

When solar patterns are incorporated in the design of houses, the outdoors comes so close that one is not always certain where it ends and indoors begins. One can observe the drama of the solstice without leaving home. You not only sense where you are in the universe but you are able to tell time without a clock and the seasons of the year without a calendar.

The Shakers understood the importance of natural, sun-filled spaces. They called sunlight "heavenly light" and all of the buildings in their farming community in Hancock, Massachusetts were designed so that they glowed "with the light and splendor of a summer day" all year long. The hen house was no exception. Light filled and warmed by a passive solar heating system, it surely housed a flock as contented as those of us who live in solar homes today.

# An Inner City
# Youth Design Program
# in Washington, D.C.

## ELIZABETH LEIGH

Is it possible to use an innovative urban design course to simultaneously improve the city and motivate and inspire young, at-risk teenagers from low-income neighborhoods?

That daunting challenge is currently being taken on by the City Vision Program in Washington, D.C. Developed as an outreach effort for 10- to 14-year-olds by the new National Building Museum, our course goes much further than a typical design class, digging deeply into the students' experience and expanding their imagination and competence by constantly challenging them to "think differently."

By introducing students to unfamiliar neighborhoods – and, via guest historians, to the unfamiliar in their own neighborhoods – and by posing penetrating values questions, the course helps students take responsibility for not only what they might design in the future but also what they are likely to do in many other ways leading to an involved, creative citizenship.

The first few sessions of the 14-week course focus on general concepts of design. Defining design as "a method of going from the existing to the preferred," we focus on problem-solving rather than on pure aesthetics. For example, students are asked to describe an object verbally by revealing only its physical characteristics while leaving out its function or similarity to other objects. The teacher then tries to draw it based on the description.

At the end of the first day, students are asked to design an egg container that will protect an egg from a two-story fall. For materials they are limited to a rubber band and an 11- by 17- inch sheet of paper. Adding to the difficulty of this problem, the students are required to write out step-by-step instructions to construct the container. The instructions are then exchanged; each

*Based on a slide show and paper by Elizabeth Leigh, architect and faculty member of the City Vision Program, Washington. D.C., USA. She lives in the Shaw Neighborhood where she works on a city revitalization program.*

student must build another's container. After completion, students and faculty walk upstairs and throw the eggs into the courtyard; usually, two or three eggs survive the impact. (Hint: success seems to hinge upon the use of wings to slow the descent as well as a cushioning mechanism to avoid breakage.)

Next, students work on technical skills such as model making, drawing and photography, and then move on to team-building skills and conflict resolution techniques. The final team-building activity, a program called "If I Had a Hammer," teaches what it's like to be on a real job site. The students have to wear hard hats and use electric screwdrivers as they work in four color-coded teams to assemble a small prefabricated house right in public view in the Great Hall of the National Building Museum. There is a color-coded map for putting the house together and each team is responsible for one side. One of the main lessons of the program is that if you can't learn how to compromise you're not going to succeed with a house or a playground or any other complicated project.

In the fourth session the group begins encountering urban realities. Working in groups of eight, with two faculty members, the students undertake field trips into their own communities. Asked to first record observations with sketches, photos, notes and collected artifacts from the environment, they are also challenged by the requirement to do a "blindfold walk" through familiar streets to determine how different things seem without the use of the primary sense. (The students also go with faculty to neighborhoods they do not know, an exercise which hones their observation skills and, more importantly, helps them compare and contrast conditions in different parts of the city.) In addition, the students conduct street interviews to gather opinions about the community from other folks who live there, and they also receive neighborhood historic information through walking tours and slide presentations by community groups and Washington's historic society.

Through a series of brainstorming exercises, they come to a consensus decision as to what neighborhood

Courtesy of City Vision Program

Model building: young inner city students learn design basics at the National Building Museum, Washington, D.C.

problem should be addressed. They then develop solutions which are supported through drawings and models. On the final day the solutions are presented to a panel of design professionals and community leaders in the auditorium of the museum.

City Visions is open to any junior high school student of the District of Columbia at no charge – generally about three dozen participate each semester. I've been involved with City Visions for three years, and my student group last year was particularly lively. Calling themselves the "Architeens," they first turned their attention to trash in their neighborhood and came up with an idea they called "incentive trash cans" – one shaped like a basketball hoop, another wired to say "Thank you!" when litter is thrown in. They then looked to identify and tackle a problem in their neighborhood, a lower-middle-income community only two miles from the U.S. Capitol.

Three of the community's major issues, according to interviews with neighborhood residents, were home-lessness, lack of day-care facilities and the lack of an outdoor market. The students decided to work on all three, focusing first on the homelessness. To do so they hit upon renovating an abandoned school as a shelter.

Through maps, drawings, slides and a model, they envisioned a building whose third floor would contain single-occupancy rooms around bathing and lounge areas. The lower floors had a cafeteria plus rooms for classes, job training, health and day-care facilities. In the schoolyard the students designed stalls for a weekend open-air market and car wash. As required by the course, the students also gave thought to financing mechanisms for their super shelter; they decided that federal funds would be supplemented by profits from the cafeteria, the day care, the car wash and the market.

Another group of students, working on a stretch of commercial buildings which has not yet recovered from the devastation of the 1968 riot, conceptualized an arcade they called Virtual World. The profits from the arcade, they decided, would be placed into a loan fund for redeveloping businesses on the street. Another project involved renovating the old, vacant Children's Hospital complex as a community mall—including a police substation, a basketball court, an ice skating rink, clinics, day-care, a radio station, shopping, dance classes, a restaurant, and facilities for homeless.

While most of the projects are conducted more for the students than for the public at large, at least one – the renovation of the large, dilapidated John F. Kennedy Playground in the city's inner-city Shaw neighborhood – is actually coming to fruition. Under the leadership of the Shaw Neighborhood Improvement Commission, funds have been raised for the playground, and some of the City Vision ideas will be incorporated in the redesign.

Inspired by City Vision, several other cities are emulating it. City Vision itself is fashioning an advanced curriculum for course graduates, one which would look at more complicated design, rebuilding and advocacy issues. Personally, I would like to see two tracks for different age groups – the current, vision-oriented course for middle-schoolers and a more hands-on course for high schoolers.

City Vision is a multi-disciplinary approach aimed at making cities work. It teaches critical thinking and problem-solving skills. It heightens young people's awareness of the community and the city at large, making them more valuable citizens. It teaches graphic skills as a method of communication. And it teaches how to work with people of different backgrounds. It's not yet talking about ecology per se but it's gradually inculcating the concept of sustainability.

If we are ever to reach the goal of creating ecological cities, we'll need lots of programs like City Vision to train and inspire the next generation of urban designers.

# Production of Combustible Briquettes from Household Refuse

## BAYE SAMBA LÔ

One of the most frequently practiced methods for struggling against deforestation is active reforestation. However, in the tropics, human beings destroy ten times as many forests as are planted, for a variety of reasons, including the following: to provide cattle with food, to prepare lands for agriculture, to get firewood and produce charcoal. The latter practice causes the greatest loss of forest cover; every year, around 50,000 acres of forest disappear because of wood consumption in Senegal.

In order to stop this trend, the government will one day be forced to forbid the conversion of forest to fuelwood. People need a substitute fuel, and we propose the use of household organic refuse to create combustible briquettes.

Charcoal has become preferable to firewood and is increasingly being used in African villages. Compared to more modern energy sources like electricity and gas, charcoal can be sold at a lower price. Though butane gas is more widely used in towns, most people – including wealthier groups – prefer to use charcoal. It seems that traditional social values indicate a strong belief that a higher quality of food is achieved by using charcoal.

A study based upon a sorting of several samples of household refuse collected in the Dakar area revealed that the components and proportions of the materials which make up the refuse reflect the socio-economic realities of Dakar's population. For example, we have found that the refuse generated by people in wealthier areas contains much more packaging materials and less organic matter. Approximately half of the collected wastes can be utilized for producing briquettes. We have determined that we could collect 365,000 tons of waste in Dakar each year. The remainder of the refuse comprises non-fuels (sand, stones, glass, etc.) that can be recycled or thrown away, as it is inert material.

The briquette production process follows these steps: 1.) Fuel elements are sorted, cut and pounded into bits.

2.) Then this prepared fuel material is mixed with water to half its volume. The liquid mixture is stirred to ensure even distribution of materials, especially paper and cardboard, throughout the volume. After mixing, the material must be doughy enough to facilitate the solidifying process. 3.) Filling the mould is next. A perforated plate is placed at the bottom of the mould that is then filled with the mixture. 4.) Solidifying the mixture is accomplished by placing a lid onto the mould, then pressing hard upon the levers against the lid. Water is expelled through the perforated bottom. 5.) Removing the briquette is accomplished by removing the lid, turning the mould upside down and pushing the bottom plate. 6.) The briquette can then be left to dry in the sun, or be placed into a kiln.

The heat content of the original refuse briquettes proved to be comparable to that of charcoal, though the briquettes burned faster than charcoal and gave off smoke. By minimizing the amount of paper, however, drying more thoroughly, and increasing the percentage of organic waste in a briquette, especially wood fuel, we could improve the rate of combustion. This should work well because the required volume of paper is at most 10%.

In Senegal, most cooking is done with charcoal.

---

*Samba Lô is an engineer working and living in the Dakar area.*

If we bring the volume of woodfuel in a briquette to 50%, we would curb today's rate of deforestation by 90% if we assume that waste briquettes are used as a direct replacement for charcoal.

At the same time, we can slow down the burning rate of fuel by using a burner that controls the air intake. A "ban ak suuf" stove is one example of such a device. Further studies will complete the feasibility assessment of this project.

The problem of the availability of fuel has been examined in the eyes of the calorific yield and the quantity of wastes which can be collected every year. The study shows that if we substitute coal with refuse briquettes, the needs of the people of Dakar can be satisfied beyond 100%.

Two possibilities of fuel production have been studied: manual and industrial. The manually operated briquette maker is made from a modest amount of steel. The handling is very easy and can be done by two operators. An alternative semi-automatic hydraulic machine saves considerable time and labor, and other technologies could be developed.

The briquettes themselves are convenient units to work with. They do not make the hands dirty and do not produce a foul odor. Once dry, they do not rapidly degrade upon getting wet, and they store efficiently. Given the briquettes' energy content, their use in industry seems promising, especially in the production of steam in power generation stations, though additional studies will be needed to explore the possibilities of this application.

Susan Felter

A street vendor in Yoff sells hot peanuts cooked over a charcoal brazier.

# Natural Wastewater Treatment Systems

## HARRIET HILL

Karil Daniels

Harriet Hill

Modern society produces lots of dirty water, which is rather hopelessly referred to as wastewater. To "reclaim" this water, we have engineered expensive treatment plants that often require lots of energy to operate, and produce large amounts of sludge which is difficult to dispose. But over the last thirty years, there has been a growing interest in constructing systems that contain natural components to help treat wastewater. These "natural treatment systems" can range from simple oxidation ponds, to constructed freshwater marshes, to highly engineered greenhouse systems that use many of the same plants, microbes, snails and fish that occur in natural wetlands to purify wastewater. Some natural systems are used to treat raw wastewater, but most are constructed to further cleanse pretreated wastewater. Natural treatment systems are often cheaper than conventional methods, produce less sludge, are usually easier to maintain, and are always more aesthetic. While it is true that natural systems may require considerably more land area than conventional treatment systems, their low energy requirements and desirable benefits like wildlife habitat creation can compensate for the higher space requirements even in densely populated areas.

Of course, natural aquatic ecosystems have been receiving and incidentally treating human wastewater for millennia. The sewage fed fisheries of East Calcutta, India which have been receiving urban wastewater for over 100 years are an example. However, when the rate of nutrient delivery to natural wetland ecosystems is increased substantially, changes in plant species composition usually occur. The most common replacement species is the cattail which can invade shrub bogs that receive wastewater. Reduced plant diversity has also been documented in freshwater marshes and tropical wetlands that receive wastewater.

*Harriet Hill is a Life Scientist with the United States Environmental Protection Agency, San Francisco, California, USA.*

Natural wetlands ecosystems also have many other important functions, such as recreational opportunities, flood control, bank stabilization, and recoverable resources such as fish and wildlife by way of the habitat it creates. Yet they continue to be drained and developed or otherwise impacted at an alarming rate. Because of the detrimental effects of wastewater on these highly threatened ecosystems, and because these systems can be artificially simulated, natural wetlands should not be used for wastewater treatment.

Worldwide, approximately 1000 artificial wastewater treatment wetlands modeled on processes in natural wetlands are now being operated. Such systems are now being used to treat almost every kind of wastewater including domestic and municipal sewage, storm water runoff, commercial and industrial wastewater, agricultural wastewater such as field and feedlot runoff, and acid mine drainage. I will present three types of water-based waste treatment systems here. These all feature macrophytes (plants large enough to be seen, that is, not microscopic) and their associated microbial (microscopic) communities:

First, floating plant ponds. These are constructed ponds which are planted with water hyacinth or duckweed. Floating plant ponds are similar in concept to constructed marshes, except that they are populated with floating plants like water hyacinth instead of "emergent" plants like cattails that are rooted in soil or some medium like gravel and emerge over the water into the air. The water in floating plant ponds is typically deeper than wetland systems, ranging from 0.5 to 1.8 m (1.6 to 6.0 ft). The plants are often frequently harvested to remove pollutants.

Second, we have constructed marshes. There are two basic kinds: subsurface flow and free water surface marshes, both featuring emergents. In subsurface flow systems, the wastewater flows beneath the surface, through a porous medium, usually gravel. These are usually planted with one or two emergent species such as cattail, bulrush or reed. The gravel provides surface area for microbial attachment, chemical reactions, and adsorption.

Free water surface systems more closely simulate most natural marshes, in that the wastewater flows on the surface through the emergent vegetation that has either been planted or established naturally. The water depths are shallow, typically from 0.1 to 0.6 m (4 inches to 2 feet). Free water surface systems are usually lined with relatively impermeable bottom soils or a subsurface barrier.

We have another category of constructed water treatment system using natural elements called "Living Machines," a term coined by ecological systems designer John Todd. This is a highly engineered system of several designs that contains elements of the above two systems and other components. One version, a facility in Providence, Rhode Island, treats an average daily flow of 9,000 gallons of effluent. It consists of translucent circular tanks that are mechanically aerated, as well as two simulated marshes at the midpoint and end. The tanks and marshes are located in a greenhouse and are filled with plants, algae, bacteria, invertebrates and fish. The operators seeded these artificial ecosystems by collecting animals and plants from over a dozen natural environments, including a Cape Cod salt marsh, southern New England streams, and a vernal pond; but they also added micro-organisms from the activated sludges in the adjacent Providence sewage treatment facility. The identity and subsequent fate of most of the introduced species in the system has not been determined by the operators, says Todd, but after a number of generations, the survivors are obviously the ones that "work."

Natural systems treat waste water by stabilizing, transforming or removing pollutants. Organic waste and suspended solids are removed through settling, filtration and/or decomposition by microbes. Nitrogen is removed in a complex process in which microbes break down large protein molecules into ammonia, other microbes oxidize ammonia to form nitrites and nitrates, and particular bacteria transform these compounds into harmless nitrogen gas. Phosphorus and metals are removed by precipitation and adsorption or are taken up by plants. Organic compounds may be volatilized, adsorbed, or biodegraded. Pathogens such as viruses and bacteria are entrapped and filtered out into bottom sediments, consumed by invertebrate predators, or die off naturally.

Plants facilitate these physical, chemical and microbial processes by providing surface area for the growth of microbes (mainly bacteria or fungi), filtering out pollutants and providing a carbon source for denitrifying bacteria. If the plants are rooted below the water, they help stabilize the bed surface of the wetland system and provide wind protection which quiets the water and prevents settled material from resuspending. Plants also shade out nuisance algae, oxygenate soil and water, insulate from temperature extremes, and take up nutrients and other pollutants.

How successful are these systems? If conditions are favorable, the removal efficiencies for natural water-based treatment systems can be quite high. However, pollutant removal efficiencies for constructed wetlands systems vary widely and nutrient removal can be problematic, especially for ammonia (a toxic form of nitrogen) and phosphorus. Living Machines remove ammonia well because they are artificially aerated and well-stocked with plants, and thus there is plenty of oxygen and substrate for nitrifying bacteria. None of the systems do particularly well at removing phosphorous, though intensive harvesting of the floating plant systems improves phosphorous removal.

For constructed wetlands, removal levels of metals and organic pollutants can be excellent. Water hyacinth ponds may perform comparably to constructed wetlands, but duck weed systems are not as effective. The Providence Living Machine removes high percentages of heavy metals, particularly cadmium, chromium, copper, silver and zinc. Pathogen removal in constructed wetlands

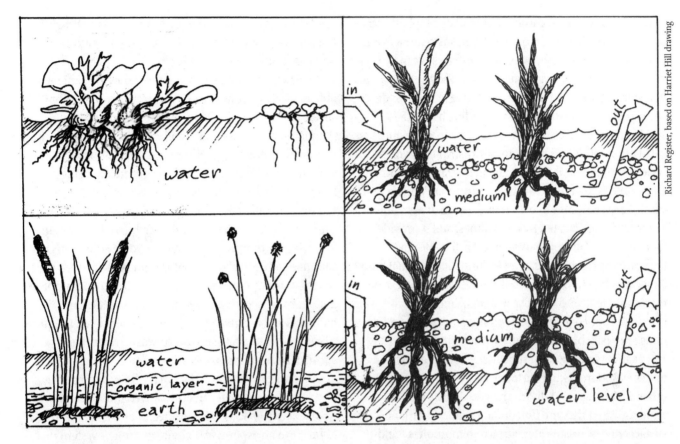

Richard Register, based on Harriet Hill drawing

Four types of constructed "natural" wastewater treatment systems. Upper left: a floating system with water hyacinth on the left and duck weed on the right. Lower left: anchored emergent plants, cattail on the left and bullrush on the right. Upper right: free surface-water constructed wetland with plants anchored in a granular medium. Lower right, subsurface-flow constructed wetland with water inside and below the surface level of the rooting medium. In both systems on the right, treated water is drawn out from near the bottom of the system (see "out" arrows above).

systems and Living Machines is very efficient, but final disinfection is still generally required to meet discharge limits.

There are advantages and disadvantages to each system. In floating plant pond systems, water hyacinths are very effective at removing most pollutants, mainly because they grow quickly and have an extensive root system as large as 120 cm or 47 inches long, providing lots of surface area for bacterial attachment and for adsorption/precipitation of heavy metals and trace organic pollutants like pesticides. They provide habitat for many other life forms such as fungi, predators, and filter feeders that help remove pollutants, and they are a good carbon source for denitrifying bacteria. Water hyacinths can effectively treat raw wastewater if it is detained long enough.

They have their disadvantages too. Water hyacinths are a tropical American species that can be a serious invasive nuisance in many waterways; Lake Victoria in Tanzania, and bodies of water in Southern United States have become destructively entangled by these plants. Also, harvesting is needed for efficient phosphorous removal. Mosquitoes are attracted and dying plants can cause unpleasant odor, requiring locations remote from human habitation. Evapotranspiration losses can be very high in arid areas, sludge must be removed from the pond periodically, and the plants are sensitive to cold conditions and create a relatively poor wildlife habitat.

Duckweed effectively removes organic waste and suspended solids and it is naturally distributed worldwide and therefore is usually not an invasive nuisance. Duckweed is extremely fast-growing and can double its surface area coverage in only four days. Its mats cannot generally be penetrated by mosquitoes, and it provides food for water birds and other aquatic life. When harvested, it can be used as a very nutritious animal food. However, duckweed is generally less effective at treatment than water hyacinth because it is small and lacks an extensive root zone for microbes and adsorption and its mats are easily blown around by wind.

Marsh systems, both subsurface flow treatment marshes and free water surface treatment systems, have other advantages and disadvantages. Subsurface flow systems, when they are functioning properly with no surface flow of wastewater, attract no mosquitoes and produce no odor. A smaller land area is required than with free water surface wetlands because the gravel provides a greater volume of treatment surface area. But they often create surface flow of wastewater due to clogging of porous media. They may produce poor treatment results if the roots of the emergent plants don't penetrate to the bottom of the gravel. Their phosphorous removal capacity is limited and usually they are poor wildlife habitat because there is little or no free water and low plant diversity. In addition the gravel bed may be expensive to obtain and install.

There are more than 130 subsurface flow systems now being operated in Denmark. Many treat septic tank effluent from clusters of two to ten rural houses and are vegetated with reeds only. A subsurface flow system in Camp Hill, England treats septic tank effluent from an institution of 60 residents where flowering plants as well

as emergents are used to create a garden-like setting. The final treated water is exposed at the end of the system in an open pond.

Free water surface treatment marshes are generally able to treat larger amounts of wastewater than subsurface flow systems. One system in Egypt is being planned that will treat 263 million gallons a day. Significant aquatic habitat can be created with free water surface systems. The litter mat, which is continually being formed by detritus, removes phosphorus, yet sludge removal and plant harvesting are usually not necessary. These free water surface marshes are poor, however, in removing ammonia nitrogen under some conditions and they need a relatively remote location because open water results in mosquitoes and decomposition creates unpleasant odors. Also, large areas of land are needed and, if habitat is an objective, free water surface marshes require a higher level of pretreatment of waste water than subsurface flow systems.

A marsh in Arcata, California, USA is perhaps the best known wastewater treatment wetlands in the world. Arcata, a city of 15,000 in northern California, was ready to buy into an expensive regional treatment system that would pipe the treated wastewater to the ocean. But town residents, including several professors from the local university, proposed a cheaper constructed wetlands system that would improve fish and wildlife habitat, and help restore a nearby urban waterfront. The wetlands proponents successfully countered political pressure to buy a large energy-intensive conventional treatment plant, and instead got state approval to construct 16.2 hectares (40 acres) of marshes to treat approximately 8.7 million liters (2.3 million gallons) of wastewater per day.

The wastewater is pretreated in a primary clarifier and oxidation ponds prior to entering the marsh treatment system, after which it flows through enhancement marshes and ultimately, to Humboldt Bay on the Pacific Ocean. The wastewater treatment marshes and downstream enhancement marshes support over 200 species

Harriet Hill

A constructed "natural" water treatment marsh in England. These water treatment facilities make excellent habitat for many plants and animals "and they are always more esthetic" than conventional treatment plants.

of resident and migratory birds, and are widely used by local residents and visitors for education and recreation.

A similar free water surface marsh in Lakeland, Florida, USA contains 3458 hectares (1400 acres) of constructed wetlands and provides advanced treatment of the domestic wastewater of 80,000 people – 53 million liters (or 14 million gallons) per day. This marsh was built in an abandoned phosphate mine in 1987. It also houses large breeding rookeries of wood storks, white pelicans, egrets, herons and many other water birds.

Looking at Living Machines, we see that they consistently deliver a high level of treatment for most contaminants and usually meet advanced wastewater standards for removal of organic material, suspended solids, ammonia nitrogen and heavy metals. Their short detention time – treatment is accomplished in approxi-

mately four days while detention times in floating plant and marsh systems may be considerably longer – is an advantage, reducing the required total capacity and acreage of the plant. Commercial recovery of resources is good, because the system is contained and relatively accessible. Recoverable materials include nursery plants, flowers, bait fish and heavy metals. It has been noticed that the behavior of organisms in the system can serve as an early warning system of malfunction to the operator. For example, if toxic loads accidentally enter the Providence Living Machine, snails quickly exit the water and move to the overhanging plants, which alerts the operator that something is amiss. Organisms such as snails, fish and shellfish feed on sludge, reducing disposal volumes.

Harriet Hill

The 40 acre (16.2 hectare) marsh at Arcata, California is one of the best known in the world. It is a popular site for bird watching and the newly-elected (November 1996) Green Party majority City Council is studying it as a potential source for drinking water.

Living Machines tend to be more expensive to build and operate than other constructed "natural" systems and conventional treatment plants for volume of service delivered. The sustainability of their performance is unclear. Designs of later systems have become more elaborate and more similar to conventional treatment plants. Significant supplemental aeration is required. To date some regulatory agency staff have been uneasy with these systems because the contaminant removal mecha-

nisms have not yet been fully documented. These complaints, however, may ameliorate with time.

The designers of Living Machines have been especially sensitive to local ecology. In order to avoid introducing alien species into natural ecosystems, they have populated Living Machines strictly with local endemic species, or species that cannot live beyond the confines of the system. For example, tropical species are used in New England systems because if they escaped, they would be unable to survive the winter.

To choose the most appropriate natural wastewater treatment system then, designers must look at their specific needs and situation. For example, if land costs are too high to create a constructed marsh system, it may be desirable to retrofit existing oxidation ponds with water hyacinth or some other floating plant species to improve the performance of a conventional treatment system. In another location a free water surface treatment marsh can become a great recreational asset, as with the Arcata Marsh in California, where the public has access to the "enhancement marshes," the diverse habitat that provides final polishing of the wastewater.

Some experts say the ideal conditions for water quality enhancement in completely natural wetland ecosystems include a well-developed and diverse vegetation community and topographic complexity and vegetation. Engineers and designers of constructed wetlands systems should investigate whether increasing the diversity of design components produces better treatment results. The design of more topographic and botanical diversity into constructed wetlands systems is now generally recommended to improve nitrogen removal. Clearly, Living Machine operators believe that the exploitation of a great diversity of plants and animals is key to their high level of performance.

# Environmental Aspects of Renewing Old Housing in Oslo, Norway

## KIRSTEN KOLSTAD

Kirsten Kolstad

As in many other European cities, housing areas built after the industrial revolution in Oslo need renewed development. Even in this small capital city of only 600,000 inhabitants, these areas have been decaying for years. This housing mainly consist of units in four or five story apartments built around the end of last century, with small backyards containing some small private enterprises such as carpentry, crafts and auto mechanics shops. The buildings are frequently in bad condition and the dwellings are often very small. Most of them lack bathrooms and proper outdoor areas.

There are several main objectives for public assisted housing renewal in Oslo. We provide bathroom facilities for each dwelling, increase the size of living space for families, establish gardens and playgrounds in backyards and make it economical for the residents to continue renting their apartments after the redevelopment process is completed. Around 11,000 dwelling units have been renewed in the last fifteen years, with all residents encouraged to participate in the process. For the most part, actual renovation is carried out by the private owners, among them some housing cooperatives. Loans and grants are given by the State Bank of Housing through the municipality and guided by a set of complex objectives including historic preservation and making the entire process and the end product as environmentally safe as possible.

Recently we undertook the remodeling of one particular apartment house, Sverdrups gate 22 (22 Sverdrups Street), as a pilot project demonstrating special environmental features. In June of 1995 the one hundred year old building of 24 small dwellings emerged from the renovation process with 16 larger units and a much reduced impact on resources and environment. Due to years of low occupancy, though there were fewer units

*Kirsten Kolstad is an architect and permaculture advocate in the Planning Department of the Municipality of Oslo, Norway.*

after renovation, more people were housed in the 16 apartments than in the 24 earlier.

Urban renewal implies reusing city areas; renewal with ecological aims implies a conscious attitude in all aspects of the renewal process, with the purpose of reducing the use of resources and minimizing pollution, in both the building process and in future management and maintenance.

We carried out several environmental measures at Sverdrups gate 22. Solar heating was one. Even in latitudes as far north as Norway's, the sun can make a significant contribution to heating houses and domestic water. So far it has not been much utilized, mainly because Norwegian hydropower electricity has been very cheap. The demand for electricity is increasing, however, and conservation efforts are becoming clearly relevant.

In Sverdrups gate 22 a solar collector system was installed on the roof facing southeast. It is used for heating domestic water and bathroom floors, covers 64 square meters of the roof and is estimated to give about the equivalent of 14,000 kilowatt hours a year in heat energy. Sixty percent of the construction costs (about US $22,000 for the project) was covered by grants. The costs for these systems have recently been considerably reduced by using plastics in the collector construction instead of aluminum and we estimate that the plastic parts will last longer than the aluminum parts.

In addition to ordinary insulation measures – which are quite extreme in Norway with double and triple glazed windows, for example – every dwelling at Sverdrups gate 22 has a central system for regulating temperature in every room for day and night throughout the week. Energy is saved by reducing temperature during nights and when the residents are away. The extra cost of installing this system – approximately $700 – is expected to be saved in three or four years. Indoor lighting in common areas is of a low energy type and has automatic on/off switches, which, so far, is unusual in Norway.

Water supplies in Norway mainly come from surface water. Oslo has had a sufficient supply to date, but demand is constantly increasing. Our pilot project conserves water in several ways; conventional ones such as water saving showers, taps and toilets, and, in four of our flats, more novel ones such as using rain water for flushing toilets. Water that flows off rooftops toward the courtyards is also used for gardens and children's play areas. These measures proved unexpectedly helpful: this year saw a severe lack of precipitation in Oslo.

Another area of ecological concern has to do with building materials and components. In Norway, the usual process of rehabilitating buildings is first to strip all old building materials and installations from the inside and then bring in new ones. In Sverdrups gate 22 all existing items were considered for possible reuse. Old doors and some windows were reused. About 400 square meters of original wooden floors were sanded, rubbed and oiled, saving the use of new boards. Our estimated savings: $15,000. Old bricks were reused where old openings in the walls were to be closed, wooden panels from walls torn out of the attic were reused in other parts of the building and some toilets were repaired and reused. Thus, in addition to the direct savings of not having to buy new materials, transportation and dumping costs were completely eliminated for these items, implying obvious benefits to forests, mineral deposits and land where construction wastes are usually dumped.

Then, some of the components of the building that were not reused on site were sold to private parties for rejuvenation and later sale. Leftovers and waste products like cut ends of wood and scraps of metal were also separated and even wrapped in some of the packaging that came with the new materials and components. Separating the waste made it possible to package it better, thus reducing volume and number of containers and thus reducing dumping fees as well as making recycling easier. The importance of keeping a tidy building site became

evident and the construction firm reported it experienced less absence through sickness and better work progress. From 1996 on, new building regulations in Oslo demand similar care in recycling and disposing of the wastes of rehabilitation projects.

Indoor climate has become a growing concern – for the workers as well as the residents who move in after the job is done. A tidy and clean building site facilitates here too, especially in regard to dust and leftovers of the building process hidden in the construction. A central vacuum system, with outlets – or rather inlets – for each apartment, simplifies cleaning. But the main emphasis has been on bringing non-toxic materials onto the site in the first place. Unfortunately, though we secured the consultation of a technical institute to study certain materials and surface treatments such as stains, paints and varnishes, it has proven difficult to find non-toxic treatments that also have a satisfactory quality and price.

Kitchen, domestic and garden waste are supposed to be separated into four fractions: paper, glass, organic waste for composting and a residual which goes to the central incinerator of Oslo. To facilitate the separation, containers for paper and glass were placed in the common yard for the residents of the apartment house. The compost facilities at Sverdrups gate 22 are the first provided for this kind of housing in the inner city. This year, 1996, inaugurates the program and compost from the household and garden wastes should be providing fertilizer later in the year for the yard and balconies. This will also reduce the amount of ordinary waste resulting in lower collection and dumping fees.

We used vegetation extensively in backyards, balconies, facades and terraces. In the attic, there is a room with windows to the north in which seedlings can be grown. The idea was to increase biological diversity, improve the local climate and contribute to the general well-being of the residents. We favored edible berries and herbs. Also in the attic, to give neighbors a pleasant place to meet one another, a common room with a terrace was constructed.

Oslo Planning Department

A group of designers and residents meet with Kirsten (center, arms folded) in a rooftop greenhouse at one of Oslo's old housing rehabilitation sites.

A presupposition was that the experimental project was not to burden the households financially. Extra costs for special features in some degree were covered either by grants, by cost-saving measures or by lower costs over time. The public grants which covered some of the extra expenses of the special features designed for environmental benefit, planning, administration, information and evaluation amounted to about $150,000 or slightly less than $10,000 per remodeled unit.

Sverdrups gate 22 is owned by a housing cooperative organized in an association of cooperatives which was responsible for the rehabilitation. The obligations connected to the project were transferred to the cooperative after the rehabilitation was finished. Now the cooperative must use, maintain and monitor the environmentally oriented features for five years and must report on these features every year. Technical measures will also be evaluated by technical experts, and the architects will publish a report on the rehabilitation process.

What else? Around Oslo there are several community gardens recently established in the urban renewal areas. These sites are either leftover areas or sites for other projects that are temporarily postponed. They would otherwise be used for car parking or illegal dumping. Short term grants have been available to help turn these open parcels into gardens for three to five years. A few years ago a small city farm was opened in the inner city and another "green" project is making the banks of Oslo's Aker River into a public park. The urban renewal areas are also provided with remodeled Slow Streets.

What next? As a result of the experiences of Sverdrups gate 22, a new program is being planned to expand the approach to other old housing rehabilitation projects. Similar grants and loans will be necessary for the different measures if they are to be attractive to the owners and tenants.

The baobab tree, common on the savanna in much of West Africa.

*As people have forgotten their roots in nature, so too have cities, towns and even the newer villages forgotten their foundations in the skin of the planet. Now to regain that balance with nature, we will need to understand what nature needs from us, so that we can redesign our communities from those ever so concrete roots on up, for everyone's benefit, including the plants and animals of Earth.*

# 11. Nature

Nature is as big as the universe through all time and space, matter and energy. Nature is as small as atoms and nagging little notions that cause one to abstain from insecticides on a favorite plant because bees will die, as well as the aphids. When it comes to cities, towns and villages, understanding their relationship with nature is as enormous a task as understanding nature herself – and as small as deciding whether or not to recycle that bottle.

Where to start on this subject is probably back in the chapter on theory, in which Paul Faulstich looks at patterns in nature and civilization simultaneously. This chapter, too, would have been a "natural" place for his contribution. But the fundamental thinking about the built community in nature he provided needed to be stated early on.

"The city in nature" is just beginning to be explored. You will shortly read about the need to preserve large areas of wilderness unmolested by humans in Michael Vandeman's contribution, with some explanations about the way natural species function. He calls for cities *not* being built in important nature areas, and for highways that tunnel under wildlife corridors – and that's a big start in beginning to withdraw cities, from locations destructive to nature. In Andrew Ventner's and Seydina Sylla's work, you will see how existing nature preserves are managed for the benefit of wildlife and humans simultaneously. Sylla speaks of placing nature preserves, for their powerful educational potential, close to population centers.

But in none of these talks will we see the interaction of city and nature fleshed out in real detail – much work remains. In fact, this is the place to lay down the challenge to Ecocity 4 and the Ecocity Movement in general: We need a new science and art of ecological city building and maintenance – and this new discipline must be founded upon the city's relationship with nature. That's why many of us call this movement an *eco*city movement, because of the all-embracing and all-supporting importance of ecology, our living context, our foundation. We need more answers here.

Is the pedestrian city the "natural" phenomenon it appears to be – in the glaring light of the failures of the city built for machines? How does the "nature" of the city relate to its "natural" environment," meaning the ecological networks of life and the flows of living and non-living (rain, wind, sun, minerals...) resources in bioregions, continents and the whole biosphere? If cities have a role to play in the evolution of consciousness, creativity and conscience in the universe, as Paolo Soleri, Tielhard de Chardin, Ian McHarg, Thomas Berry, Anne Whiston Spirn and others suggest, what is their healthy anatomy and metabolism? Can this be described and visualized in ecocity maps, and strategized in clear "steps to an ecology of economy" as proposed in Chapter Three? Can we really rebuild our civilization in balance with nature?

The next three presenters reveal a quiet heroic struggle to better understand our relationship with a healthy natural environment. They give us a foundation for a kind of "applied natural history," a system of designing and building, bringing our cultural ways into harmony with nature's.

# Wildlife and the Ecocity

Shinichi Uematsu

Michael Vandeman at Mount Okue, Kyushu, Japan

## MICHAEL VANDEMAN

### The Problem

We are losing, worldwide, about 100 species per day. Habitat loss is at the top of every list of the primary reasons why species have become extinct or are in danger of becoming extinct. Destruction of habitat by paving it, turning it into farms, golf courses, housing developments, or urban parks is not the only way that an area can become untenable as habitat. Anything that makes it unattractive or unavailable to a given species causes habitat loss. Many animals simply will not tolerate the presence of humans. The grizzly bear and mountain lion are just two examples. The grizzly needs a huge territory, can smell and hear a human being from a great distance, and will avoid going near a road.

If we are to preserve the other species with which we share the Earth, we need to set aside large, interconnected areas of "pure" habitat that are entirely off limits to humans. Our idea of what constitutes viable habitat is not important; what matters is how the wildlife who live there think. When a road is built through an area, many species will not cross it, even though they are physically capable of doing so. For example, a bird that prefers dense forest may be afraid to cross such an open area where it may be vulnerable to attack by predators. The result is a fragmentation of habitat: a portion of pre-ferred foods, mates and other resources has become effectively unavailable. It can lead to extinction. Small populations isolated by human intrusion can easily be wiped out by a fire or other disaster.

In 4 million years of human evolution, there has never been an area off limits to humans – an area which we deliberately choose not to enter so that the species that live there can flourish unmolested. There are places called "wildlife sanctuaries," where human recreation,

*Michael Vandeman is an environmental activist and citizen planner active in many issues in the San Francisco Bay Area of California. He heads the Wildlife Committee of the Sierra Club's Bay Area Chapter.*

hunting, logging, oil drilling, or even mining are usually allowed. There are a few places where only biologists and land managers are allowed. There have been "sacred" places for priests only. But to my knowledge, there has never been any place, however small, from which the human community has voluntarily excluded itself.

In other words, we assume for ourselves a right to travel anywhere that we want and we deny that same right to wildlife. As a side effect of building our cities, we have created practically impassable barriers (impassable even by us: ever try to cross a freeway on foot?) that prevent wildlife from going where they want to go. Where did we get that right? Do we really need to go everywhere and do everything that we fancy? One of our proudest moments is when we are able to go somewhere "where no human has ever gone." Maybe instead of proud, we should be deeply ashamed. Ideas of right and wrong evolve (as they should), as our knowledge of the world evolves.

Try this experiment: go on a hike, find a tree that you like, lie down under it on your back, and look up at the tree. How do you feel? Watch how the wind causes the branches and even the trunk to sway, and yet always return to where they were. Note how many other organisms live on and in the tree and enjoy what they find there. The first time I did this, I suddenly realized that this being, the tree, had lived in this one spot all its life, and was happy and content to be there. How, then, could I pass by any spot on the Earth and not fully appreciate its value, its sufficiency? Do I really need to visit every country on the Earth? Even though I love hiking, do I really need to hike every trail in the world? How could I ever again take lightly the cutting down of a tree?

We assume for ourselves a right to live. We deny that same right to other species. We think nothing of killing a tree, even a very large tree for the White House or our local shopping mall Christmas decoration. Millions of birds have been slaughtered, some driven to extinction, in order to decorate ladies' hats. Killing for survival may be necessary, but killing for anything less crucial, in light of the current biodiversity crisis, is archaic.

We assume that every individual human life is unique and priceless – that we each contribute something special and invaluable to the world. On the other hand, when we speak of preserving wildlife or biodiversity, we are usually talking about preserving only species. I have never heard any biologist admit that individuals can be important – even genetically. But don't new genes appearing for the very first time appear in a single individual? And what if such an individual were killed? What if the Christmas tree that we cut was one that contained a mutation that would allow such trees to survive global warming? I don't think it is safe to say it is OK to lose individuals, as long as some of the species survive, in zoos for example.

How did we decide – rationally, I mean – that humans are more important than other species? Our genes are 98 percent identical with those of a chimpanzee! Every other organism can do things that we can't do and many of those things are "services" that make human lives not only enjoyable, but possible. I wonder how long we would survive without nitrogen-fixing bacteria, photosynthesizing plants and organic material-decomposing organisms. I know of only one fair way to judge other species: by the same rules we apply to ourselves. That is what we call the "Golden Rule," except that we now apply it only to other humans.

We assume a right to clean air, clean water and clean food. We deny wildlife those same rights. We assume the right to eat whatever we want. We don't allow wildlife the same right if they choose, for example, one of our pets or livestock. We demand privacy in our bodies and homes. We deny wildlife that right, entering their homes and habitats, and violating their bodies with impunity whenever we wish. When we set aside land that is to be touched as little as possible, we call it a "park", meaning "a human playground." We want wildlife there but we don't go much out of our way to ensure that they con-

tinue to flourish there far into the future. One of my local park directors actually campaigned for election using the slogan "Parks Are For People!" I would like to replace the word "park" with "wildlife habitat" in order to set priorities straight.

Worldwatch just published a paper ("Eco-Justice: Linking Human Rights and the Environment") by Aaron Sachs in which he recommends that environmentalists make use of local and indigenous people to protect their environment. However, he does not talk about wildlife, and does not recognize that they may have a need and, hence, a right to be left alone, and not have humans living in their midst and "harvesting" them, even at a "subsistence" level – whatever that is. Even indigenous people have joined the market economy and have begun harvesting in "industrial" quantities.

When we aren't abusing wildlife, we are ignoring them or taking them for granted. Look at travel guides. For example, the Lonely Planet's guide to Japan contains only half a page on wildlife. They constitute the section called "Dangers and Annoyances." It seems that the only thing worth knowing about the wildlife of Japan is that you should avoid its two poisonous snakes.

Why do we treat wildlife so badly? Are we evil? I don't think so. I think there is a relatively simple explanation. The brain is optimized for efficiency: it pays attention only to what is needed for immediate survival. We learn to ignore most of the information that is available to us, so that we can focus on what is most important at the moment. We just haven't had much immediate need to focus on wildlife. And in our highly urbanized world, we think we "need" to focus on wildlife less than ever. Most people take wildlife for granted. They "have their hearts in the right place," but haven't examined the issues and don't know the facts about the biodiversity crisis. My role is to try to wake them up, gather important facts and spread information around the world. That makes some people uncomfortable – mountain bikers, hikers, or even scientists who don't want to be excluded from wildlife habitat. But my goal is simply to face the facts squarely, since I believe that is our only hope.

**What Wildlife Need**

What do wildlife need? Reed Noss, Michael Soule, Ed Grumbine, and other conservation biologists have addressed this question. Wildlife need..., just what we need (remember our 98% genetic commonalty with the chimpanzee?). They need a place to live where there is an adequate supply of food, water, potential mates, protection from predators, etc. For long-term survival, this area must be large enough to supply all of those needs, as well as contain a big enough gene pool – that is, a large enough number of individuals – that they don't get inbred and can adapt to changing circumstances such as fire or global warming. In addition to secure habitat, they need safe corridors by which to travel to the resources they need.

Most proposals addressing the question of how to preserve our biological heritage have called for a network of large, "inviolate" (no, or almost no human use) reserves, all connected by wildlife corridors, and "buffered" by land with minimal human presence. To save all of our current complement of species would probably take at least 50% of the land area of each continent. Wildlife are not as flexible as we are, and most require very specific kinds of foods, climate, or terrain. The key, of course, is not what we think is adequate, but what the species themselves want and need to thrive. Large carnivores, for example, require huge territories with minimal or no human presence in order to find an adequate supply of prey and a reasonable choice of mates.

Roads and other human facilities that cross these corridors are a problem. In order for a corridor to function as a corridor, organisms must be able to travel safely. Some species will use a tunnel under a road, but some will feel that such exposure is too risky, and won't use the tunnel. Ideally, a road should tunnel under the wildlife corridor. How frequent do these crossings need

Richard Register

Conferee Nancy Lieblich marks the location of creeks in Berkeley, California as part of an educational campaign for opening creeks. These 860 stencil images on curbs – a different "creek critter" for each of the city's twelve creeks – identify the places creeks pass under Berkeley streets, 85% buried in large pipes.

to be? Only research will tell, but since we don't have time or resources for all that research, erring on the side of caution (that is, disturbing wildlife habitat as little as possible) is the wisest approach.

**What to Do**

We need to adopt prudent guiding principles. For example, since wildlife cannot protect themselves from us, and since their requirements are much stricter than ours, they must be taken care of first. Planning for wildlife habitat areas and corridors should precede all other planning, and, the fact that a particular area has already been damaged for use as habitat – clearcut, or turned into a golf course, for example – and hence no longer functions adequately as habitat, should not be used as an excuse to "write it off" and damage it further. Wherever there is damage, we can plan for restoration.

Aaron Sachs, writing for the World Watch Institute, put it this way: "Protecting the rights of the most vulnerable members of our society is perhaps the best way we have of protecting the right of future generations to inherit a planet that is still worth inhabiting." Of course, he, the entire Worldwatch Institute staff and the long list of other reviewers of his paper all failed to notice that "the most vulnerable members of our society" are wildlife! If such an erudite bunch can commit an oversight like this, I guess the rest of us can be excused ours.

What better place to begin rectifying our abuse of wildlife than in our parks? They provide some protection to wildlife already and could provide the seeds of a full function habitat and corridor matrix. But first, as I said earlier, the word "park" should be replaced by "wildlife habitat" or "wilderness." Next, we should remove, as much as possible, all human artifacts – buildings, manicured areas, exotic plants and animals, parking lots, most trails excepting short, educational "nature trails," and, above all, roads. Perhaps we could allow bathrooms and drinking fountains, at least in the most heavily used areas, for health reasons. Of course, all mechanical forms of transportation, such as jeeps and bicycles, would be forbidden. We should make an exception for wheelchairs, but other than wheelchair-accessible nature trails, I would not want to see wildlife habitat sacrificed for any humans, even the disabled – the most disabled human is still better off than many species of wildlife, which are going extinct. This is the most humane way to reduce human impacts on the parks – not exclude people, but just make the wilderness a bit harder to get to.

In line with our laissez-faire approach to park management, natural processes should be allowed,

Richard Register

Salamander tunnel under highway outside Amherst, Massachusetts. Four-inch fences angle out into the forest to funnel migrating amphibians safely under the road.

including fire. Those who put their homes right next to a park should also bear whatever risks that entails, including fire, falling trees, and, if food in the park gets scarce, predation of pets. They should also carefully watch their children, the way animal mothers do. Park managers in my area, under pressure from nearby homeowners, have been trying to "fireproof" the parks, and are destroying wildlife habitat in the process. In short, many parks should be allowed to revert to wilderness, and wilderness should be a place that we enter rarely, reverently, and on its own terms.

In addition to parks, which are ideal places for people to learn about nature and sustainable living, there need to be "core" areas that belong entirely to wildlife, and are off limits to people. Every region should have such areas, partly for the sake of wildlife, but also for their educational value: their very existence would silently and continuously teach people an enormous amount about biology and the ethical treatment of wildlife, just as architecture nonverbally communicates how the designer wants us to think about something or someone. Of course, an area that we never intend to visit, does not need to be depicted on the map. So I propose that we blank out these areas on every map, and designate them simply "terra incognita," or some such expression that conveys the proper respect. I call this "de-mapping." Maps, by facilitating human access, can be just as dangerous as roads. Like roads, they are a two-edged sword, giving equal access to good and evil.

Psychologists tell us that children learn most of what they will ever know by the age of six. This education, of course, is mostly nonverbal. They also develop an attachment to their surroundings called "imprinting." I believe, therefore, that every infant, soon after it meets its mother and father, should be taken to experience wilderness. We need wilderness. Most of the intelligence (information) in the universe is stored there (in strings of DNA); we crave intelligence (information). Surprise lives there. Also delight. Our brains thrive on complexity; too much simplicity literally puts them to sleep (or gives rise to hallucinations, as in sensory deprivation). Wilderness – for example, swamps, coral reefs, and rain forests – contains most of the world's complexity. We pride outselves on our ability to empathize with others. Let's demonstrate it, with wildlife. Let's try treating individuals of other species just as we want to be treated – the ecological golden rule. If we are as skilled at communication as we believe we are, let's communicate with other species, and find out directly from them how they want to be treated (but, come on, don't we really know already?).

**The Ecocity**

My definition of an environmentalist is someone who gives top priority to the welfare of wildlife. I believe that everything else that normally characterizes an environmentalist follows from that premise. For example, if we provide clean air for wildlife, we will automatically do so for ourselves, as well.

An ecocity is the same: it is simply a city that puts wildlife first – a city that lives by the above priority scheme. From this axiom, you can derive the theorems that flesh out the ecocity. I am not going to attempt to describe that exactly. For one thing, a detailed design for any town would require detailed information about local indigenous species. The best that I can do, I think, is offer some tentative suggestions. I am sure that you, applying the principles I have presented, can come up with much better ideas.

First, recognize wildlife as equal citizens of the community, with the same rights as humans, but, being virtually defenseless against human technology, a special case with special protections, in a sense "above" or "beyond" the laws governing human interactions with each other. These wildlife rights should be overridden only when necessary (for good cause).

Second, minimize pavement! Let's start removing all unnecessary pavement while we still have enough oil (fuel) left to do it! It's no fun doing it by hand. Third, our natural "resources" are sacred, having intrinsic rights separate from human utility. Wetlands are sacred. None

should be destroyed or covered up. All creeks should be liberated from their above ground and underground prisons (i.e., pipes, culverts, concrete channels, etc.). Water should not be polluted. It belongs to all species. Soil is sacred. It takes eons to create. It can be moved, but should never be destroyed. It should not be polluted. It belongs to all species. Air is sacred. It should not be polluted. It belongs to all species. Finally, since wildlife cannot protect itself from us, it should be accorded top priority. Planning for preserving wildlife should precede all other planning.

These principles should be codified and implemented by the United Nations and concurrently studied and adopted by every government agency, business, and private organization.

All over the world, humans are fighting, invariably over some resource that both parties covet. Wouldn't it be wonderful if both sides could be distracted for a moment from their selfish pursuits in order to come to the aid of a third party of even greater need? The best gift that we can receive from wildlife is to give humans who are fighting selfishly among themselves and creating wars a way to work together to forget our differences. In the past, people have seen children starving and said, "my God, the children are in trouble, let's do something." The same goes even more so for wildlife. And the logical vehicle for this process is the ecocity movement.

Monkey in Senegal's Siné Saloum nature preserve.

# Interface of Traditional Communities and Game Preserves

## ANDREW VENTNER

Ecotourism is a big buzzword. What does it mean? To me it means sharing a country's lifestyle with people. Much is conservation, but it is cultural, too. It is what we are doing during this conference week. This week is ecotourism. Much of what we are doing is learning about people. And many sorts of projects are being explored. There are community development projects looking at short term problems, water supply and things like that, and there are very important research projects which look into the future. In them we are working together and figuring where we are going for the future.

At Kruger National Park, we have recently established forums to begin such exchanges with peoples living near the park. For the first time in eighty years, local community leaders and Kruger Park personnel have sat down and started talking to each other. Some of these rangers have spoken with people beforehand, but it has been the "kick in the door and gun to the head" type of approach. This is a new approach. We have started talking, learning together, and designing a new future based on new management practices and policies.

The forums involve three core groups: community members, conservation authority people, and then the external organizations, which are very important – government and non-government organizations that have expertise and skills that can bring knowledge in, but they do not run the process. The idea is for the community members and the conservation people to form one community. They work together. They now start deter-

*Andrew Ventner is the consulting biologist for the Kruger National Park, largest in South Africa. Due to only partial recording of his presentation and failure to elicit more information from him by mail, we have relatively few words here, but ones with lessons worth sharing. He spoke first of the old policies of park management under Apartheid under which the traditional peoples, black peoples, were extricated from the game preserve. Then he went on to describe a new philosophy and policy under the present Mandela government.*

mining the future of that park. There is a whole range of different options that they are looking at. The focus at the moment is planning, working together.

One tool we use to involve people in planning is the Participatory Rural Appraisal (PRA) workshop. It is about working together. Instead of going in with a questionnaire and asking people questions like, "Where do you live, how many in your household, what do you eat and drink," it involves working together and learning to share experiences. It is participation from the bottom up.

There are opportunities for human resource development, skills training, either formal or informal. A simple example: in the park there are people who can drive a bulldozer. That is a marketable skill, but the people who live in the communities do not have access to those skills. It is very simple for the park and interested people to get together to share those skills.

There are projects near Kruger Park that the park has become involved with. One is a nursery that grows trees appropriate for planting in the area and provides a source of seed. Another is a permaculture garden. Our colleagues at EcoLink spearhead projects like water management systems in places where water is a big problem. We are involved in recycling projects, including finding appropriate and creative uses for litter. We have contributed teacher training in biology and mathematics.

Many of the people living outside the park have already been alienated from the park. The one priority that the teachers in these communities ask for is training in environmental education.

We are working with local people to use the park to stimulate the local economy through small business development. For example, Kruger Park generates about $50 million every year. Before we began working there, the park was buying about $1 million worth of curios every year for resale to tourists. Of those curios, less than $4,000 worth came from the local communities. The rest came from Morroco, Kenya and other countries. We are starting to shift that. There is a vision here.

# Safeguarding Senegal's Biodiversity – The Plan for Protected and Peripheral Zones

## SEYDINA ISSA SYLLA

The degradation of biodiversity is a phenomenon that is not recent. It went on for years before people became conscious of the intrinsic value of biodiversity on environmental, genetic, social, economic, scientific, educational, cultural, recreational and aesthetic grounds.

Human responsibility for the regression of biodiversity – plant and animal species and whole habitats – leaves us with the moral homework to watch over safeguards for our remaining natural resources. In this regard, Senegal ratified the Convention on Biodiversity in 1994.

It is to safeguard biodiversity and the durable, or sustainable, utilization of biological resources of Senegal that we are involved in developing national strategies in respect to this convention. To implement a national strategy demands above all the development of a national monograph, or document, that permits us to judge the environmental gains and losses, and thus to better target our priorities. Certainly the production of this document requires knowledge and experience in these areas and Senegal has already asked a team of experts to begin the task. The monograph will address protected places and their peripheral zones, after achieving a logical frame-work in which different actors (NGOs, grass roots groups, research institutions, and government institutions) participate. Thus rational management of bio-resources can be established and different tools and activities can be applied. At the same time, this document will aid in establishing national strategies supported by international experts.

Historically, the phenomenon of impoverishment of biodiversity started to be felt in Senegal in the first half of this century. The initial causes of this phenomenon were attributed to the expansion of cultures which greatly increased private income detrimental to natural wooded areas that permitted the fauna to exist. This biological

impoverishment intensified in the second half of the century with the pursuit of agricultural practices that consumed space and resources. The increasing number of domestic livestock inflicted great damage on pasture ecosystems, especially in the Sahel. Climatic assaults characterized by two decades of drought were brought about by practices of development that made a clean sweep of cultural values and their ability to accommodate nature. Ecological catastrophes in the 1970s finished by offending national and international public opinion.

However, the effects were judged as a result of deforestation, the remedy was identified as reforestation by way of aid, subsidies and loans. Large quantities of resources were consecrated to reforestation and were considered the only means to contain the phenomenon of spreading desertification.

After twenty years of reforestation politics, it was realized that deforestation persisted. The state then established a network of protected areas consisting of representative samples of different biological realms (with the exception of the Sahel area).

We will discuss two sites. The first is in the humid coastal zone of estuaries and deltas, which have multiple habitats, and which are especially noteworthy for serving migrating sub-Arctic birds. It represents an area of 20,000 hectares (50,000 acres) of islands, marshlands, lakes, mangroves, forests and 50,000 hectares (125,000 acres) of marine environment offshore.

The second area constitutes a park of 913,000 hectares (2,250,000 acres) of savanna, Niokolo Koba Park, created to preserve the last representatives of the great land fauna of our country. These areas were established with the involvement of organizations like the UN World Alliance for the Conservation of Nature and the World Wildlife Fund. They were also established as part of the country's commitment to conventions it signed, including the World Heritage Network of Man and Biosphere, UNESCO African Convention (Algiers, 1968), and the Convention on Biodiversity and Climatic Change (Rio, 1992).

*Conservateur-Commandant Seydina Issa Sylla is head of the National Parks System of Senegal. Synopsized from the paper he delivered in an abbreviated form at the conference.*

Bill Mastin

White pelicans join 300 other species of birds at the Djoudj National Park in Senegal, one of the three most important bird sanctuaries in the world. From November to April, some three milion birds migrating south from Europe pass through this area.

In the middle of the 1980s the political powers recognized that the conservationists were right – the state of urgency was real. Up to about that time, consulting with the nearby people was not felt to be necessary. However, in the last ten years the old classic politics yielded to more local involvement and rational management. By 1993, the Interministerial Council for the National Management launched its strategy for safeguarding biodiversity. Now grassroots organizations collaborate with our government in making policy for stewarding sustainable biological resources.

Gains and losses are both evident since the mid-1980s. The network of park and reserve environments that were protected were soon reinhabited or in process. Across the board, biodiversity is now maintained and there is great improvement outside the parks as well. Medium and large fauna were saved in Niokolo Koba and the aesthetic quality of the sites was preserved or restored adding to the potential for tourism. The result was demonstrative but imperfect. The dry Sahelian area – actually one half of the country – is not part of this network. One must also note that certain species are in danger, principally the elephant, the lycaon, the Derby deer, the panther, the lamantin and the chimpanzee.

In the last two or three years there has been some international cooperation, however. Germany financed the installation of the biological station of Bjoudi (1993) that maintains a welcome for scientific research. Other support has come from the Financial Foundation of Dutch Cooperation, the European Development Fund and the country of France, which is helping with a financial study for the management of the biological reserve of Niokolo Koba – results presently pending.

With respect to the Biodiversity Convention, the combination of projects and activities will be carried out within the framework of an action plan that must respond to the following principles: 1.) involvement of national and local NGOs, 2.) women's participation, 3.) use of the traditions practiced by native populations and their involvement in the management of the environment, 4.) provision of better living, social and economic conditions for local populations, 5.) cooperation of research institutions and sharing of information and 6.) popularization and exchange of pertinent technologies.

Our strategy incorporates the following guidelines, which were developed at the Interministerial Council of 1993 responding to the Convention on Biodiversity: 1.) establish maintenance policy and practices supporting biodiversity in the protected areas, 2.) define supporting activities for the local population in sustainable natural resource management where the resources still permit it, and for tourism and ecotourism in these protected spaces and their peripheries, such as village hunting, game ranching, traditional fishing, management and development of meat "in the sticks," bamboo cultivation, beekeeping for honey, pharmacology, etc., 3.) plan for the reconstitution of biodiversity where it was damaged (particularly in the Sahel zone) through rehabilitation of genetic stocks in permanent natural surroundings and through the establishment and development of long-lasting management activities and 4.) generally, respect and maintain harmony with the landscape.

It is evident that such a program cannot be made without increased public awareness in different sectors of society, at both the regional and national levels. Large numbers of people need to know about environmental problems and their profitable socio-economic solutions.

Also, the training of employees in technology appropriate to the preservation, rehabilitation and development of natural resources in various ecosystems is essential to this strategy. This will necessitate an exchange of know-how with people in similar places that have experience with these appropriate technologies, especially in regard to fauna. For example, we plan such an exchange with South Africa and Zimbabwe.

With respect to specific regions, our country divides naturally into three principal geographic and ecological areas: wet delta, coastal and estuary, and the Sahel savanna zones.

Situated in the wet delta, estuaries and coastal zones, the Parc National des Oiseaux du Djoudj (National Bird Park of Djoudj, a World Heritage Site) and the Parc National de Delta du Saloum must be regarded with particular attention because their plants and animals need a better economic symbiosis with the neighboring rural communities.

We are studying the idea of creating a management authority for water in the natural wet zones of the Senegalese delta. This authority would permit a coordinated management of water between various entities. It should be noted here that installation of necessary works to control the waters – to maintain nesting sites in the bird park at Djoudj, for example – are urgent. The studies preparing the management plan for Djoudj, which are being realized under financing by Holland and lead by the International Union of Conservation of Nature, take account of the local population's aspirations.

The annual declining numbers of water birds in the Senegal Delta is a serious present concern. The solution looks like it will need to be complex and therefore the management of these protected areas necessitates efficient scientific research.

With regard to the public awareness of the environment, the small protected areas close to the towns such as the National Park of the Madeleine Islands and the Popenguine Nature Reserve, must take on their proper importance. The latter provides instructive material examples for the rehabilitation of natural vegetation put to the test. The Fathala Forest is an ideal site for bringing together wild herbivore Senegalese animals, city dwellers and coastal tourist clients.

In the savanna ecosystems, our primary concern is the biology of the Niokolo Koba Park where only a part of the park conserves the last representative of the great fauna of the country. The strategy there is to rehabilitate its infrastructure, then supply the park in a manner that permits it to play a positive role in the touchy process of breeding. It's an experimental approach. Meantime, we need to put the peripheral areas to good use.

The acquired experience there will permit us to establish a true eco-management approach at other selected rural communities close to the park, threading together village hunting, game ranching, ecotourism, meat "in the sticks" etc. These activities must be economically viable and also must be led by and for the local populations represented by feminist groups, youth groups, NGOs, and all others. Parallel to these manage-

Djoudj National Park in Northern Senegal, near the Mauritanian border.

Bill Mastin

ment activities is the promotion of bamboo, honey and so on, for both economic and ecological purposes.

It has only been since 1990 that the Niokolo Park benefited from the annual count of the big and medium fauna and some of the small fauna such as guinea fowl. This was accomplished thanks to the aid of the French Institute for Scientific Research for Development and Cooperation and the Canadian-Senegalese Partnership Fund. These counts, which permit following the population trends, are of great importance for designing management plans and decision making of all sorts. Management of fires and the mastery of invading species such as certain mimosa trees that encroach on ponds are important to study.

The Sahelian steppes have been rapidly degraded by livestock and drilling for their water. Much of the native fauna, with the exception of some ostriches and red foreheaded gazelles, are now extinct, and much of the covering vegetation is in a state of ruin. The damage caused by the saturation of the area with livestock will probably require us to ban further drillings in the area and to close certain existing wells – to the great displeasure of the livestock breeders.

The area's protected stream environments cover a small fraction of this landscape, but these are biologically very important and must be protected. Our Parks Department should be the driving force for the rehabilitation of all Sahelian biodiversity even though little of it is directly under our jurisdiction. For that, the National Parks Department of Senegal and other institutions concerned with rehabilitation must have redefined roles.

The strategy for rehabilitation of the Sahelian biodiversity we call eco-management. Along with reintroducing wild Sahelian species in these areas, such as breeding gazelles and ostriches, we are protecting livestock. This is in deference to the local people's present economic need and because there is not enough remaining fauna to politically support repopulation of native species at this time. The reintroduction or "reloading" of these animals in the area, and the upkeep of pasture, is

Susan Felter

Heron in flight in Senegal's Siné-Saloum Delta, near the Gambian border.

for the collective benefit of both fauna and livestock. The plan must be made in collaboration with different players of which the principal ones are the breeders. This must be a long term draw on the benefits of rehabilitation for rational management. Proper value must be placed on the natural resources for tourism, study of fauna, pharmacology, etc. Again, it will help to orient breeders, women's groups, youth groups, local NGOs, and all the other local groups in the formulation of these projects and in the research for funds. And we need to help them develop their own activities for redevelopment, management and the promotion of biodiversity.

The traditional West African village as depicted on a wall mural in Yoff.

Future small ecocity designed for the Australian Outback by Paul Downton.

*An infinite number of kinds of cities are possible,*
*including terrible ones. Choosing only good ones,*
*still an infinite number are possible.*

## 12. Conclusions

The way we editors looked at it, "this conference ain't over 'till the report's published." But it will continue as memories in the minds and hearts of all the conferees as well, and as a new experience in a widening circle of readers of this book, *and* in the following conferences of the International Ecocity Conference series. And so, with the ink drying on these pages we actually see what they call at school graduation ceremonies, a commencement, a beginning, not just a finish of this phase of the journey toward the ecocity.

Here we have our own best effort at answering the question, "What can we do to build an ecocity civilization?" Our answer, crafted at Ecocity 2 and refined at Ecocity 3, is The International Ecological Rebuilding Program. (We know there will be modification, fine tuning....) If we assumed the responsibility – *somebody* should do it – to think through what we need to do to rebuild human society in balance with nature, it might look something like this. The Marshall Plan for rebuilding after the Second World War provides some precedents for such an enterprise. International treaties on biodiversity, climate and other natural "resources" are steps toward just such a rebuilding program. And so are the many initiatives described in this book, for building ecovillages, and reviving inner cities, downtowns and neighborhood centers, for new transit lines and urban growth boundaries, for appropriate technologies and mixed-use solar buildings, for waterway restorations and wildlife habitat expansions, and a lot more.

But the cornerstone, the clear idea of the physical, strong and friendly built ecological community is not yet there in government policy, business strategy or citizen consensus.

But all this *can be*, and we offer it here. If only governments, by the hundreds, businesses by the thousands and people by the millions would sign on. Perhaps they will! It's all up to you.

In the closing days of the conference and in all those attendant bestowals of *terenga*, and at the closing ceremonies of the conference itself, many memorable words were spoken – some recorded here.

Other words, other ideas, other images linger on, among real physical conference outcomes, some healthy and growing, some stillborn. We contemplate a few of them here, consult our recollections of the conference for those pearls of wisdom (hoping we have found some of the more important ones) then invite ourselves to dream on, as all people must, before the creative work begins again.

# The International Ecological Rebuilding Program

*This document was introduced and rewritten in open committee at the Third International Ecocity Conference. A similar version was adopted by the Second International Ecocity Conference in Adelaide, Australia. To date, no government, from city to national, has adopted the "Program" but as you will see next, in "Outcomes," Joan Bokaer is working on a strategy now.*

## Introduction

The International Ecological Rebuilding Program aspires to influence humanity's city-building and village-building activities everywhere. We have a crisis in our present way of building our communities – and we believe that, in writing and adopting this document, the conference has attempted a long step toward a solution.

Strategically, this program aspires to lead to changes in policies, life ways, ideas and values. As an article of faith backed by much evidence and experience, the authors believe that the built communities of our species can benefit both humanity and nature and need not assault either. It is a matter of how the built community is conceived, built and maintained, and that, we believe, is clarified here.

## Preamble

Under the laws of nature and of humanity, we, adopting *The International Ecological Rebuilding Program,* declare that the time has come to build our communities in just consideration of an ecological healthy world and in compassion for all its living beings. This program aspires to encourage responsible design, planning and building and take them to a new level of integration: integration of social, economic and ecological goals and values. It seeks to help clarify and launch an international enterprise that will inspire the building of cities, towns and villages that will take part in the cycles of nature like other living organisms, in a way that brings health and thriving to all.

Since 1972 at Stockholm, when the United Nations launched a series of conferences addressing human impact on the environment, both natural and human-built, urban population has more than doubled and the impacts of human activity on the biosphere have accelerated. We are now facing loss of natural habitat and whole species and world-wide atmospheric build up of carbon dioxide threatening to change climates and raise oceans. Efforts to establish more equitable distribution of resources to disadvantaged human groups, efforts to conserve energy and to preserve species are now written into treaties and laws and yet humanity continues developing in highly unsustainable patterns with enormously destructive impact on nature.

A major part of the solution – and one that is little recognized – has to do with the layout and arrangement of the built community, from village scale to city.

Cities, towns and villages have become ever more dependent upon high energy use, long commute distance and machine-based transportation. Yet, both very recent and ancient traditional examples of building have contributed to many successful pedestrian community centers and whole quarters of cities and whole villages. It is not too early, in 1996, in examining these successes, to conclude that designing and planning communities for great diversity of functions at close proximity, together with major efforts to restore nature and agriculture, indicate a far healthier future than the one prepared by designing and planning for automobile dependence, sprawl development, large highways and high energy consumption. The ecovillages and ecocities that can result from the effort to build communities on the human measure – the chiefly pedestrian community – would function at very high levels of efficiency in recycling, energy and resources use. All society and all of nature would be the beneficiaries.

The aim of this document, *The International Ecological Rebuilding Program,* issued on January 12 of 1996, at the conclusion of the Third International Ecocity Conference, held in Yoff and Dakar, Senegal, then, is to urge a new perspective and strategy in the process human beings employ in building their communities. We the conference participants believe the time has come to examine the natural principles behind healthy community organization and construction. We believe that such an examination, currently in progress in the Ecocity Conferences and in the work of many pioneers around the world, is now ready for presentation to governments, businesses, Non-Governmental Organizations and individuals around the world. We believe the time has come to institute an International Ecological Rebuilding Program and that the following principles are a good beginning at developing such a program.

## The International Ecological Rebuilding Program

1. Reshape cities and withdraw from sprawl.
   A. Restore, establish and preserve village and city centers of great economic and cultural diversity with people living, working and taking part in the full range of community activity in a small radius.
   B. Write zoning laws that:
      1. designate existing centers of diverse activity as pedestrian centers,
      2. identify areas for restoration of nature and agriculture.
2. Revitalize traditional villages, small towns and rural areas.
   A. Create economic bases and social and cultural incentives for people to thrive in their activities.
   B. Protect traditional villages from invasive urban sprawl.
3. Restore natural environments and productive ecosystems.
   A. Restore, establish and preserve natural areas and farm and garden lands surrounding cities.
   B. Restore, establish and preserve wildlife corridors for ranging and migratory patterns of animals and provide continuity for plant communities.
   C. Promote reforestation and reverse desertification.
4. Design cities for energy conservation and recycling.
   A. Design efficient, compact villages and cities with circular, integrated systems in which the waste of one process becomes the raw material for another. Reuse and recycle to the fullest extent possible.
   B. Establish policies that encourage vigorous investment in renewable energy and recycling technologies.
   C. Minimize the production of non-recyclable waste and end the practice of exporting non-recyclable waste.
5. Build foot, bicycle, and public transportation. Create a transportation infrastructure that makes walking, cycling and public transportation safe, convenient and accessible.
6. End automobile subsidies. Assess cars, gasoline, roads, bridges, and parking lots, and collect taxes, tolls and fees that reflect their true cost to the environment and society. Spend the revenue on ecocity rebuilding.
7. Provide strong economic incentives to businesses for the ecological rebuilding effort.
   A. Establish policies that encourage investment in ecological village and city rebuilding.
   B. Tax activities that work against ecologically healthy development.
   C. Develop ample retraining programs – millions of jobs will be needed to reshape our built environment and restore our natural environment.
8. Create a government agency at all levels – city, state, and national – for ecological development.
   A. In this agency, explore the ecocity vision.
   B. Consolidate mutually supportive or interdependent functions facilitated by government (such as transportation, energy and land use) that are presently separated in bureaucracies.
   C. Facilitate actual construction and restoration projects to build ecocities and ecovillages.

*Adopted by the Attendees of the Third International Ecocity Conference, Yoff and Dakar, Senegal, January 12, 1996.*

# Closing Statements

## LIZ WALKER, ADJI ARAME THIAW, FATOU LO, BIRAME THIAM, FLORENCE CHERRY, THOMAS MACK

The closing day of the Third International Ecocity Conference was as beautiful, sunny and spirited as the opening day.  Again the eloquent hosts assembled and celebrated their guests.  The International Ecological Rebuilding Program, just preceding these words, was discussed and endorsed.  And many words were spoken and some written for our parting meditations.  We'd look back on these words in our homeward bound flying machines, some headed toward the worst blizzard in 500 years on the North American east coast, or in the sandy streets, to the sound of prayers called from the mosques five times a day.  Again, groups of people rose from the audience to sing to the speakers.  But finally we said good-bye.  These are some of our last thoughts from the conference itself.

### Liz Walker

*Liz Walker is co-director with Joan Bokaer, of EcoVillage at Ithaca.  She walked with Joan and hundreds of others in the march across the United States called the Walk for a Livable World that Joan organized in 1990.  Here she answers a question on camera, recorded by our videographer Karil Daniels.*

We have a hundred people who have come from all parts of the world to Yoff, to Dakar, to find out about the culture here.  And people are learning about a way of living together in community which is very, very different from the modern industrial world.  It's a real eye opener.

Perhaps the symbol that was most profound to me was when we went to the traditional village of Ngor and we saw the tree that had been there – I believe it's a dabouliya tree – it had been there for six hundred years and the elders gather there every Friday and resolve the problems of the village at the base of that tree.  And they have propped up the branches of the tree, which are very spread out – they propped up those great branches with pillars.  I love to see that image of nature being very much a part of the people's lives and very much respected just as the elders and their wisdom are respected.

Adji Arame Thiaw

### Adji Arame Thiaw

*Arame Thiaw was head of APECSY's team of hostesses and guides.*

What I enjoyed most about the conference was the incredible value and credence people from around the world, despite all their twentieth century ecovillage technologies and concepts, gave to our traditional way of life.  To me the major teaching from the conference is that the people of Yoff, despite the cultural, social and political changes pervading throughout the world, still have a priceless quality of life; Yoff is a community where cooperation for the benefit of the whole has always been a given, where governance for centuries has been based on consensus decision making.  And this, I think, is something many in the rest of the world only dream of.

### Fatou Lô

*Fatou Lô was one of our many gracious hostesses before, during and after the conference.*

During the conference, I got in touch with many other people concerned about preserving the environment.  In one word, my knowledge increased a lot.  According to me, the only thing that the traditional village can teach other people is its life in community.  The village is not rich enough, but it leads a quiet life.  People are very nice and they help each other.  Here in Yoff we have big families.  The family doesn't mean only the married couple and the cat but all the relatives.  Yoff is the kindness itself.

### Birame Thiam

*Birame Thiam was speaker Harriet Hill's guide in Yoff.*

The Third International Ecocity Conference has been of great importance to me.  It has taken place at a time in which the world is dominated by individualism and materialism. In fact, the second millennium is tinged with scientific and technical progress which have made man slave of his own discoveries. Man has been brought so high in materialism that he has no longer the time to contemplate the marvels that nature offers him. And so, the ecocity conference means a return, or a reconciliation of man with himself, of man with his environment and his nature. The conference has allowed me to understand that no people can gather all elements necessary for this balance. You can be morally and culturally opulent but materially bankrupt or be materially opulent but morally and culturally bankrupt. So for a well balanced world, there must be complementarity between peoples.

I consider the great wisdom of Yoff to be its ability to maintain its parallel social organization based on a traditional legacy, with its head, the Djaraf, the Ndey Dji Reews, the Freeys, etc. alongside the state with its President, ministers, police and so on. In fact this traditional form of organization does in no case hinder the state organization. Quite the contrary, it helps for socialization and better communication between the members of the society through familial ceremonies.

Fatou Lô

## Florence Cherry

*Florence Cherry is an anthropologist with the faculty in the Department of Human Development and Family Studies, College of Human Ecology at Cornell University, Ithaca, New York. She chaired panels for the conference and spoke at the closing day ceremonies.*

Florence Cherry, on the right, with speaker Elizabeth Leigh.

Salaam. It has been my great pleasure and privilege to be a delegate from the United States to the Third International Conference on Ecocities and Ecovillages.

The past four days of this conference have been for me another step in my personal and academic development. I hope that these four days will not only make me a better anthropologist but a better citizen of my own country and the world. For me the concept of ecocities and ecovillages has become an integral part of my personal and academic vocabulary and I know these concepts will play a role in how I interact with the environment.

The opportunity to live and begin to learn the traditional culture of a Senegalese village has only convinced me of what I have always known, that what the tradition of Mother Africa has to offer to the developed nations of the world is profoundly important.

I want to take this time to thank the conveners of this conference for not only choosing Senegal, West Africa for the site of this conference but most especially for selecting the village of Yoff. A more fitting place could not have been chosen.

Since this is my first trip home to Mother Africa, the village of Yoff could not... *(she breaks into tears and the audience claps loudly and appreciatively)*... have been a

Karil Daniels

Thomas Mack

more fitting place to pass this week. I feel like the prodigal daughter who has come home to my elders, my mothers and my sisters and brothers of the village of Yoff. And while it is necessary for me to leave you for a short time, you can be sure that your prodigal daughter will be home again to Yoff very soon.

## Thomas Mack

*(Our speaker on permaculture.)*

We are all creatures of this Earth, born of the same soil and vested with the breath of life given to us by one God. The words human and humus have the same roots. As we touch on the Earth, it is the responsibility of each and every one of us to express our gratitude for the gift of life and to work diligently to protect the environment and the commonwealth heritage of the Earth's natural resources.

Today I would like to present to you a gift which is symbolic of the miracle of life, which science and technology with all its fascinating toys are powerless to duplicate. *(Reaching into an African wooden bowl, Thomas produces several ears of Indian corn.)*

These seeds were grown by traditional native American farmers from a village in New Mexico which is 800 years old. We have here every color of humanity: black, red, yellow and white. They are a symbol of the unity of life and of our brotherhood. These are the seeds of peace and today may they inspire us to think of future generations and remind us to care for the Garden of Eden that we have been given by God. May they also symbolize the hope that this conference has inspired. This gathering is a seed for a future in which we may all live in abundance and harmony with our mother planet Earth.

*At this point, Thomas presents Serigne with the bowl of multi-colored corn and again, people rise in the audience, singing, drumming and swaying. Presently the meeting is declared closed and we file out into the fresh warm night, three hours behind schedule. At the main mosque, the Khalifa says a prayer for our safe return, and as we leave we realize that a few cherished faces are already missing and on their way to the airport.*

# Outcomes

## RICHARD REGISTER

Three hundred people's lives were changed, many of them, deeply. Professional and personal connections were made, in the usual way of conferences, and future work will be enriched by the experience.

The International Ecological Rebuilding Program was adopted. Thus a call for and a commitment to substantial rebuilding of cities and villages on ecological and socially equitable grounds was made, and a call for recognizing and building on the experience of traditional villages was emphasized. Should a real Ecological Rebuilding Program be adopted by governments and supported by businesses and individuals, the crisis in the way humanity builds could turn into one of our greatest successes – a cornerstone for a healthy future in full respect of cultural traditions would be set in place. If connected directly or indirectly with Ecocity 3, this would constitute a stunningly positive outcome.

The International Ecological Rebuilding Program took only small steps immediately after the conference, however. It is very ambitious and its prescriptions are strong medicine for a society addicted to cars, oil and sprawl development and becoming more so rapidly. A side note in these last days of editing this book: in the fall of 1996 the World Health Organization (WHO) announced that high stress diseases, especially heart and lung diseases, and car accidents "will become" the number one and number two causes of death around the world by the year 2020, surpassing communicable diseases. Number two in the whole world for car accidents?! That is a mind-boggling quantity of violence and death to attribute to a transportation device, to something that would become obsolete and unimportant in two or three generations if we were to build our settlements in consciousness of ecological principles and for "access by proximity" rather than access by motor vehicle.

Rather than roll over and let this "will become" run us over like the five axle, eighteen-wheel truck of Uncontested Fate, something as strong as the International Ecological Rebuilding Program will be required. Whether society will have the courage to face up to its structural

Yoff street scene.

(physically constructed) addiction to cars and gasoline and take the appropriate medicine is another question. It was difficult getting recognition for the "Program" at the United Nations Habitat II Conference in Istanbul, June 1996, in any case. Some progress was made there in supporting alternatives to automobiles. Compact development with mixed land uses was featured in the long list of needed changes endorsed by the United Nations in their "Action Agenda." But nothing close to a call for an ecological rebuilding program in general principle was adopted, much less *The International Ecological Rebuilding Program.* At Habitat II there was some influence felt from Ecocity 3, but it was very small.

Direct outcomes in the village of Yoff were more clear cut. A permaculture design certification course was held immediately after the conference led by our presenter Thomas Mack, and within weeks several gardens based on these ideas and techniques were flourishing in Yoff, including an expanded women's cooperative garden and a new garden in the front yard of the APECSY headquarters building. We hear that they are heeding Anders Nyquist's advice and have created a system for utilizing urine to build soil without violating Moslem prohibitions against handling body wastes. The result, according to conference host and co-convenor Serigne Mbaye Diene, is spectacular garden production. I suspect

Conference videographer Karil Daniels (with camera and earphones) at a conference celebration.

that few, if any, conferees returning to their western toilets made any such progress.

Internet connections were set up by Jon Katz of Ithaca, New York and Yvonne Andres of Carlsbad, California, working with APECSY members and installing donated computers and modems. A library on ecological technologies and ecovillages was started, and progress was made toward the designing of an ecology center.

A "Yoff Eco-Community Program" (Eco-Yoff) was established, with conference speaker Marian Zeitlin selected by APECSY and the village elders as co-director with Serigne Mbaye Diene. An "eco-arts festival" sponsored by the program is being planned for every year on the anniversary of the conference starting in 1998. "Eco-Yoff's mission," says Zeitlin in the newsletter, "is to develop a comprehensive strategy which will sustain the fishing community of earlier times, as it is absorbed into the expanding city of Dakar, and make Yoff into a sustainable urban neighborhood by the year 2020."

To meet this goal, permaculture and other ecologically-based technologies and systems will be taught through the schools. Village religious and social traditions will be documented and available through a museum. Eco-Yoff's "environmental subcomponent," says Zeitlin, "will address the erosion of the shoreline, deforestation, protection of biodiversity and soil erosion."

Marian Zeitlin continues with the most telling outcome of the conference – showing just how difficult it is to make substantial progress. "Future residents hope to develop an ecovillage, but not as a poverty project. The attractive low cost modern housing units which are starting to go up on the Extension are not ideal from the perspective of preserving traditional culture and the traditional ecological lifestyle. Unfortunately, no one markets the old ways. The new units are for single families and are spaced for car traffic, unlike Yoff's traditional communal compounds which are home to 50 or more extended family members and are joined by narrow pedestrian walkways."

In the last twenty years Yoff has doubled in size. To actually walk across it lengthwise now is quite a chore and one end seems remote from the other. To expand out into another 150 acres on one end, adding to the length, will reduce Yoff's physical accessibility, and hence, will inevitably erode its sense of unity.

Several conferees with experience in ecological planning had proposed retaining Yoff's traditional street pattern and its clustering of homes around compounds. We can see that this advice was not accepted. In the fall of 1996, the first of the single family "modern housing units" and automobile streets are under construction.

Not only the outside "advisors" at the conference, but also many people in Yoff wanted to retain their traditional housing and street patterns in the new housing. But the Municipality of Dakar has ordinances requiring wide streets for "emergency access," not considering that prevention is the better part of cure: the way of life in the pedestrian environment makes many emergencies far less likely. The World Health Organization report that said car accidents would be the number two cause of death in 2020, said that number one would be heart and stress related diseases: exactly the kind of cause of emergencies that wider streets would theoretically serve – while causing! – by bringing in cars to hit pedestrians, by creating air pollution and noise, by increasing stress levels, by dividing elite car owner from poor non-car owner, by fracturing social security, and round and round.

The more imaginative ideas for Yoff, such as the proposal to provide new housing by constructing infill buildings and additions near neighborhood centers and making them taller than today's average height while *not* expanding into the "Annex" area, saving that land for farming, were not adopted from the conference Annex workshop proposals. Reminding the Annex planners that such taller buildings have a long tradition in Africa, and in Moslem areas at that, such as in the Kasbah cities of northwest Africa, had little effect.

Thus *the* outcome that could have been the one most characteristically "ecocity" in nature, and the one with the most positive effect on village tradition, structure, agricultural land and biodiversity, was too different from the present development trajectory in that part of the world to be adopted. Remember that the High-consumption World is not rushing to adopt most ecocity recommendations either. No blame implied – it's just the way it still is; the planners of Dakar were trained in the same schools as those of the West. And remember Jeff Kenworthy's admonition: we need to throw the manuals out the window and write new ones – based on ecological principles. *Not* easily done. It is interesting to note Mayor Mamadou Diop's welcoming address to the conference again: "The city aggressively sends out its tentacles upon its traditional villages and surrounding land to asphyxiate and consume them...." As much as Serigne Mbaye Diene, head of APECSY, and Mamadou Diop, Mayor of Dakar, would like to change that modern Western development trajectory, its momentum is for the moment unstoppable.

Perhaps the most subtle and sad of all changes just after the conference is the new plan to name the streets and put numbers on houses. No longer will the guests need to talk to the villagers for guidance, nor will the villagers be needed to help the visitor find the house of the such and such family. Perhaps that kind of intimacy, that kind of personal welcoming was what really made Yoff such a friendly place. Perhaps towns just can't work without street names and house numbers after a certain population is exceeded. Maybe this is the fundamental divide between the traditional village and small city.

Joan Bokaer's evolving ecocity strategy is another outcome of Ecocity 3. As Yoff is having difficulty taking ecocity changes to the depths of its land use foundations, Joan's conference-inspired efforts to expand on ecocity work in Ithaca, New York came up against a frustrating lack of response there. Despite the substantial success of the ecovillage just outside town, despite the growing awareness of the importance of sustainable development, despite the largely successful Ecocity 3 and signs of a growing international movement, "mostly what I have been getting," she says, "is silence. I think the problem is that people don't believe what I am suggesting is possible. They don't see how to do it."

What she is actually proposing is this: First, link key places in the city of Ithaca with trolley infrastructure. Three lines could connect Cornell University, Ithaca College, downtown, two major parks, two neighborhoods, Ithaca High School, a junior high and the board of education building. Most of the city could be connected by three trolley lines. But the population density isn't sufficient to warrant trolley lines, so this problem leads to the second component of an overall set of changes: building transit villages along the trolley lines. Mixed use pedestrian "transit villages" on rail lines would help fill in parts of the city with a population that would use the trolleys, enrich downtown, and increase the tax base of the city without increasing vehicular traffic.

There are some ideal locations for these transit villages, Bokaer tells us, potentially beautiful places on waterways, one in downtown on a creek, one near the Ithaca Farmers Market on a canal. Presently they are eyesores. "In every city you can find such places amidst potential beauty and vitality, land lost to parking or run-down buildings. These are great places for in-fill."

"The third component of this plan" she says, "is to use a percentage of the cost of building the transit

Bill Mastin

House with built-in bench, Yoff.

villages to conserve land surrounding the city so that, as the population is filling in the city, precious farmland and natural areas surrounding the city are conserved. For example, if a transit village costs $200 million to build and 10% of that cost goes for land conservation, then $20 million would be available to purchase development rights on land surrounding the city, and conservation easements would be placed on that land thereby creating an urban growth boundary or greenline for the city. As population moves inward, countryside is forever conserved.

"The beauty of a small city like Ithaca is that once the three components are in place – a trolley infrastructure, transit villages, and a city surrounded by open space including family farms, ecovillages and natural areas – then people could visit and see fairly quickly how those elements all work together. They can experience the vitality and fun of pedestrian centers, the ease and convenience of a good rail system, and the beauty of a city surrounded by farms and natural areas. Visitors would experience the feeling of truly integrated planning.

"As we build models of ecological cities by transforming existing cities, at the same time we begin the process of rendering the automobile obsolete. It is not that people love their cars and long commutes so much; it's just that they don't know what they're missing. Wait until they get a taste of how wonderful life can be without cars! My job is to help people see that it's possible to transform our cities.

In addition, without policy support and investment from a much larger portion of society than is now participating in the ecocity movement, success of the trolley/development/land conservation idea would be difficult or impossible. Therefore Joan is proposing to make Ithaca the first city to adopt the International Ecological Rebuilding Program. Success here would not only make it easier to develop as she hopes by changing attitudes and policy, but would make success more likely by attracting the kind of serious tourism that already goes on in some measure in university towns like Ithaca. Ithaca would become a destination for contemplating ways we might be living in the future and experiencing the vitality of a living, relevant, pioneering experiment.

Joan Bokaer accomplishes most of the projects she undertakes. Keep an eye on this one! In fact, help it happen. Presently she is planning a planning process for Ithaca that could lead to adopting the International Ecological Rebuilding Program and to a land and rail development project based on ecological principles.

These ideas and work span into what might be happening next. But we had a more direct link in mind to publish here. We had hoped that under "Next" we could print greetings from the convener or conveners of the next one or two ecocity conferences. Cleon Ricardo dos Santos of Curitiba, Brazil, and Bente Florelius of Bergen, Norway, both present at Ecocity 3 and represented in this book, are promoting the idea of holding these events in their cities, in Curitiba probably in 1998 and in Bergen probably in 2000. But these decisions are not yet made. Stay tuned, folks. Ecocity Builders in Berkeley, the Open University for the Environment in Curitiba and the Planning Department of the City Government of Bergen will have up-dates. Other locations are possible too. The ecocity conferences live on, but the real item "Next" on the agenda is what all of us, dear ecocity builders, are going to do to help transform our communities.

# Some Summary Thoughts –
# The Village, the City and the Future

## RICHARD REGISTER

What then have we discovered? What is the wisdom of the village that can help us into the future? What can we learn from those among us who have been working to make our cities a tool for our compassionate, creative evolution as a species, which means in addition (if we extend our compassion and imagination to other species) that our communities would also become ecologically healthy? We editors have portrayed life in the village and the talks at the conference as best we could and the content will, for the most part, have to stand on its own, to be interpreted by each of you. But we would like to highlight a few points that seem especially noteworthy.

Many of the conferees were planners, designers and activists with a strong interest in the physical structure of villages, towns and cities. We immediately noticed the extraordinary peacefulness of the narrow, soft sand streets of Yoff. Observed California photographer Susan Felter, "if you fall down, you don't get hurt." Children fall and they pop up laughing, rather than crying and rubbing bloody knees and elbows. Public spaces are soft spaces here, not concrete, asphalt or cobblestone. Very interesting! No noisy cars to intimidate and run children over – or anyone else. We the visiting conference participants tended to look deeper than the obvious, but over and over it came back to us that these gentle streets were, *obviously,* one of the profound lessons in themselves. The pedestrian street, the "walking street" or the "walking city," in the words of conference speaker Jeff Kenworthy, is humanizing, and it is absolutely essential for making our cities ecologically healthy, too.

Most of us felt that the sharing and cooperation that was everywhere evident in Yoff was another powerful lesson. Where you can drop in on relatives when you are out of a job and they not only feed you but actually celebrate your presence – *there's* social security! We were aware that the villagers were going all out to be gracious, generous hosts and some of us visitors were cautious of over-romanticizing a culture with some serious problems as well as many good points. But that too kept coming back to us – it was undeniable that these people seemed exceptionally comfortable and secure, generous and warm, thoughtful and peaceful.

In rural village cultures there is often a pervasive sense of limits of material wealth. Urban Europe, largely through primogeniture and mercantile re-investment, was good at accumulating wealth – and the weapons to acquire and defend it. The rich industrial nations, calling themselves capitalist and coming to include North America and Japan, were even better at it, democratizing (somewhat) access to wealth within their own borders. Thus as vast numbers of people – about a third of the world's population – became collectively an enormous engine of acquisition and transformation of resources, into products, urban civilization became exceptionally good at accumulating wealth and power. Some of these people became excellent at patronizing art as well as war, science as well as exploitation, nutrition and medicine as well as environmental destruction. Thus (in sinfully brief outline) was created the fascinating but wildly dangerous world in which we live – dangerous for other species and many of our own. The few still control the wealth, like the families of aristocracy and royalty three hundred years ago in Europe did, and whole countries are the wealthy minority. Other whole countries are destitute and in debt, as Tukumbi Lumumba-Kasango pointed out early in the conference.

The sandy streets of Yoff – "you fall down, you don't get hurt."

Contrasting with city civilization, rural village culture has tended to not only share, but also to be very poor at the accumulation of wealth. The disparity between poverty and wealth is much less in such cultures than in systems of privilege, despotism and rampant consumerism. In fact, when those with more are not busy sharing material wealth with the ones with less, through tradition, mores, social pressure, religious stricture or the guilt of looking a deprived neighbor straight in the eye on a daily basis, they are busy burying their accumulated wealth with the dead, sacrificing it in supplications to God or otherwise making sure it doesn't accumulate *too much*. This ceiling of material wealth, this striving for balance and harmony, this *ecological* sense of limits is foreign to the High-consumption World that celebrates its most avaricious people as paragons of a new kind of virtue. Roll over Jesus, Buddha and Gandhi – here comes Morgan, Getty and Gates.

But the results are getting obvious in the collapse of living systems and change of climate. Perhaps we have a lot to learn, here in the traditional village. Graduated income tax? More than that, village experience suggests we need an income limit, a ceiling. Say, for example, nobody makes more than five or ten times what the poorest in society makes. We the editors don't know the ratio in Yoff, but we do know it displays nowhere near the disparity we see in the globalizing economy that celebrates ever wealthier individuals and corporations while numbers of people in deep poverty multiply rapidly around the world.

There are other lessons from the village. Taboos, things you just don't do, are common in traditional village cultures and are often inexplicable to outsiders. But there is a big lesson here that makes common cause with the environmentalists who say "no" to a long list of dangerous technologies and practices. Nukes are becoming taboo. Forswearing the private car should eventually be among them and the sooner the better, though as we see in many places in this book, this calls for major ecological redesign of the city.

Something particularly evident in Yoff: there is a lesson in the way the older children, as well as parents, take responsibility for the youngest, and the whole society lives with the elders everywhere present and held in high esteem. The result seems to be great independence and comfortableness in society on the part of the children and a loving communication between all ages.

What we can learn from the people who came to Yoff to talk about their designs, plans, and community projects, starts off with the similarities, rather than contrasts. It seems that the lessons they bring are, for the most part, common to both villages and cities. Some of the similarities are these: First, "access by proximity" functions for healthy physical, economic and social processes. Closely related to this, pedestrian arrangements (supported by bicycle where the streets aren't made of sand), rather than motorized vehicle systems (worst of all automobile systems), are the most human and ecologically healthy means for transportation. Closely related again, in both city and village, scatterization erodes ecology and society. Perhaps last among the big lessons, biological and ecological richness adds to the richness of culture and economy at all scales.

Ask each conferee what he or she remembers and values most about this particular conference and the answer is almost always the same: the extraordinary opportunity to live in the traditional village and get to know the people.

But there is also a deep lesson from the ecocity innovators here, consistent with the village ways we experienced, a lesson about harnessing the human creative force for a great social and ecological good. The vitality of creativity is deeply kindred to the vitality of violence and destruction. If this comment seems to land like a flying saucer in the middle of this paragraph, let's think it through. The spark of vitality in all life forms, and its channeling through different physical and behavioral patterns, is at the center of these issues and not alien at all. Creativity can transmute into something else, becoming productivity, over-productivity, glut, greed, hunger for power and finally destructivity. In fact, the frustration of the creative drive often leads directly to lashing out, as if violence would break through to new opportunities to be creative. Building cities is basically a creative activity, creating something that wasn't there before the city-building enterprise began. In fact, there is no larger human creative activity than building cities. The only trouble is that we are beginning to realize that this kind of creativity – creation of the very living structure we are part of – has transmuted into one of the most, perhaps *the* most destructive phenomenon on Earth since the great "extinction spasm" of 65 million years ago in which the dinosaurs disappeared.

It need not be so. We are now recognizing ever more forcefully that wastes can become new resources if they are recyclable following the model of biological processes, in the "ecocycles" Anders Nyquist cites. We can design with the whole systems perspective of ecology, for distances between elements of the "land-use/infrastructure" at "human scale." That is, we can build pedestrian environments. The profound lesson of the traditional village walking street is reiterated by ecocity theory and was brought forward by all of our speakers whose work goes beyond the one-building-at-a-time level to the arrangement of buildings in a whole living system.

Appropriate technology is just a start: bring on appropriate creativity! The vitality of creativity, the sense of being at home that comes with conscious and sensitive participation in building society's home, channels human activity toward health and deals with the distortion of the human impulse toward creativity and wanting to change the world. To make the built community of service to humanity and Earth at the same time, then, constitutes perhaps the largest physical enterprise conceivable that channels human vitality toward cooperation and compassion. Ecocity insights let us know how to direct creativity toward ecological health and social justice – in settlements of all scales.

After establishing strong bonds between visitors and hosts, after returning home, further communications

Most of Yoff's narrow streets open to the beach and the ocean's distant horizon.

from the young guides of Yoff to the guests, were both touching and troubling. In several letters brought to the attention of this book's editors, not only did many village correspondents express sadness in missing the guests, which the guests also felt after leaving their hosts, but some of the village youth said they were very depressed. Many wanted to come to the guests' countries as soon as possible. Some hinted at great personal troubles and a few asked indirectly for money to help solve their problems.

Around the world young people who are not being forced to move by economic forces, sometimes want to leave villages for the perceived more exciting urban world. Many get out and never look back. Very few who leave return. This impulse probably has a lot to do with the desire to live an exploring, creative life, healthy in itself, but destructively exploited by the purveyors of today's urban world on the make. As conferee Nancy Lieblich said, "for a lot of the young people of Yoff, the conference probably gave them a taste of the outside world and was extremely frustrating. The people who came from far away got to get back on the plane and take the trip home." The local people, faced with very low financial resources, can not only *not* satisfy their curiosity by buying an airline ticket themselves, they usually can't even muster the resources to go to a school of higher education. With a tiny handful of West African Francs (CFAs), they sit there in real social security, but look up at the airplanes leaving town to all points of the world.

This may be especially frustrating in Yoff: the end of the takeoff strip launches the planes directly over the western end of town. Imagine this sight which I saw with my own eyes. Looking down a quiet street with children practicing a dance step and older folks walking past a horsedrawn wagon, we can see one or two sheep wandering from place to place. In the background, behind the humble two story sand-colored houses and ancient trees that look ever so small, even fragile, an unbelievably gigantic machine, getting larger and larger, lifts into the sky. It looks larger than an ocean liner. It completely dwarfs the sleepy town, rising almost silently as its four massive engines fire out their noise away from us, as the nose rises higher and larger overhead. Suddenly the massive 747 is overhead and its shadow, as wide as the whole neighborhood, is rippling over the houses, the people, cart, horse, sheep. Thunder consumes this small world. Off to the west banks the big jet, now growing small, with more than two hundred people, each with tickets about 20 times what the villager could afford, getting smaller and smaller as the silence and distant murmur of the ocean return. Most of the visitors won't. Every day several such planes take to the air over Yoff, the international airport to the whole country.

But this problem of the allure and promise of distant places, always intoxicating for young people, if unique in its extreme for Yoff, is an underlying reality in the city - village relationship everywhere. An enormous issue rises like that airplane here: how do traditional villages and rural societies more grounded in sustainable relations with nature keep the young people home? The answer may be that you don't try to keep them from leaving. You may even help them leave. But you make both the cities and the village more honest about and responsive to the ecological realities. The village can adopt some of the creative aspects of the city, and the city some of the ecological aspects of the village. Both can become more exciting and relevant. A young person can stake out an adventure of a life in either place. Under such conditions, a civilization could be born, a civilization of both cities and villages learning from one another about directing creativity, through ecological principles and values, into a new kind of community. The village can become a place where the young are far more likely to stay or return home to after travels, education and adventure. Reverse migrations from the wealthier urban world to the village world and back – such as exemplified by the ecocity conference itself – would similarly reward the wealthy world with important new insights.

These notions ran deep beneath the surface of the projects represented at the Third International Ecocity Conference. Joining the wisdom of the ancients and the attempt to redesign our cities – the combination is like mixing oxygen and fuel, lighting a lamp to the darkness. What could be more positive, more hopeful, more promising?

Karil Daniels

# Resources

## Organizations (and Presenters)

(Marina Alberti), Department of Urban Design and
Planning, Box 355740, University of Washington,
Seattle, WA 98195-5740, USA

American Farmland Trust, 1920 N St., N.W., suite 400,
Washington, DC 20036, USA

APECSY — L'Association pour la Promotion
Economique, Culturelle et Sociale de Yoff (Serigne
Mbaye Diene), B.P. 8502, Yoff, Senegal, West Africa,
fax 221 24 24 78, apecsy@yoff.yoff.sn

Arcosanti, the Cosanti Foundation, Arcosanti, AZ
86333, USA

(Moussa Bakayoko), Direction de l'Aménagement
Urbain, Commune de Dakar, BP 186, Dakar,
Senegal

Bergen Planning Department (Bente Florelius), Bergen
Kommune, Radstuplass 5, 5017 Bergen, Norway,
fax 47 55 566330

(Janis Birkeland), The Center for Environmental
Philosophy, Planning and Design, Department of
Architecture, University of Canberra, P.O. Box 1,
Belconnen, ACT 2616, Australia, fax 61 6 201 2279,
jlb@design.canbera.edu.cu

(Lee Boon-Thong), Department of Geography,
University of Malaysia, 50603 Kuala Lumpur,
Malaysia, fax 60 3 759 5457

(Ivan Cartes), School of Architecture, University of
Nottingham, Nottingham NG7 2RD, United
Kingdom

Center for the Built Environment, 2/5, Sarat Bose Rd.,
Calcutta – 700 020, India, fax 91 33 466 0625

Cerro Gordo, Dorena Lake, Box 569, Cottage Grove, OR
97424, USA. cerrogordo@igc.apc.org

CoHousing Network, 1620 Belvedere, Berkeley, CA
94702, USA

(Amalia Cordova), Museum of Pre-Columbian Art,
Domingo Toro Herrara 1471, Nunua, Santiago,
Chile, fax 562 697 2779,
acord@uchdcc.dcc.uchile.cl

(Mamadou Diop), Mayor of Dakar, P.B. 186, Dakar,
Senegal

Earth Island Institute, 300 Broadway, San Francisco, CA
94133, USA, fax 1 (415) 788-7324

Ecocity Builders (Richard Register), 5427 Telegraph
Ave., W2, Oakland, CA 94609, USA,
fax 1 (510) 649-1817, ecocity@igc.apc.org

EcoLink (Sue Hart, Gregory Vlahakis, Elise
Mpatlanyane), Box 727, White River 1240, South
Africa, fax 27 1311 33287

Ecological Design Arts (Thomas Mack), 369
Montezuma #225, Santa Fe, New Mexico, 87501,
USA, fax 1 (505) 989-9854

The Ecological Design Institute, 10 Liberty Ship Way,
Sausalito, CA 94965, USA

EcoVillage at Ithaca (Joan Bokaer, Liz Walker), Anabel
Taylor Hall, Cornell University, Ithaca, New York,
NY 14853, USA, fax 1 (607) 255-9985,
ecovillage@cornel.edu

enda: environmental development action in the third
world, P.O. Box 337, Dakar, Senegal,
fax 221 22 26 95

(Paul Faulstich), Pitzer College, Claremont, CA 91711,
USA, mefeldman@aol.com

Findhorn Foundation, the Park, Findhorn, Forres, IV36
OTZ, Scotland, thierry@findhorn.org

Fossil Fuels Policy Action Institute and the Paving
Moratorium, P.O. Box 4347, Arcata, CA 95518,
USA, fax 1 (707) 822-7007, autofree@tidepool.com

Gaia Villages (Ross Jackson), Skyumveg 101, 7752
Snedsted, Denmark, fax 45 97 93 66 77,
gen@gaia.org

Global Fund for Women, 425 Sherman Ave., Suite 300,
Palo Alto, CA 94306-1823, USA

Global Schoolhouse/Global SchoolNet Foundation,
7040 Avenida Enunus, Carlsbad, CA 92009, USA,
fax 1 (619) 931-5934, andres@cerf.net

Healthy Cities Program of the United Nations World
Health Organizations (WHO), Len Duhl phone at
1 (510) 642-171, www.healthycities.org

(Harriet Hill), US Environmental Protection Agency,
2618 Hillegass Ave., Berkeley, CA 94704, USA,
fax 1 415 744 1078

INRECON – Central Research and Design Institute for
Comprehensive Reconstrucion of Historic Towns
(Vitali Lepski), 121293 Poklonnaiya Str., 13,
Moscow, Russia, fax 70 95 148-93-95

International Forum on Globalization, P.O. Box 12218,
San Francisco, CA 94112, USA, fax 1 (415) 771-1121

(Luis Jugo), Urb. El Encanto, Qta. Inna, Mérida,
Venezuela, fax 58 74 66271

(Jeff Kenworthy), ISTP, Murdoch University, South St., Murdoch, Perth, W.A. 6160 Australia, kenworth@central.murdoch.edu.au

(Kirsten Kolstad), Architect, Planning Department, City of Oslo, Maridalsveien 86A, 0458 Oslo, Norway

(Elizabeth Leigh), City Vision Program, National Building Museum, 1869 Mintwood Pl., NW #31, Washington, DC 20009, USA fax 1 (202) 265-8510

(Fatoumata Lelenta), Educational Specialist, AENO Palais de Justice, Dakar, Senegal

Los Angeles Eco-Village/CRESP, 3551 Whitehouse Place, Los Angeles, CA 90004, USA

(Tukumbi Lumumba-Kasango), 311 W. Sibley Hall, Cornell University, Ithaca, NY 14853, USA, fax 1 (607) 255-6681, tl25@cornell.edu

(Richard Meier), Prof. Emeritus, University of California, Department of Urban Design, 636 Colusa, Berkeley, CA 94707, USA

(Birgitta Mekibes), Bjorkallen 18, 14266 Transunp, Sweden, fax 46 8 6040942

(Anders Nyquist, architektkontor), S - 862 96 Njurunda, Sweden

Oikos Bioedilizia, Società Cooperative R.L., Via Crescini, 147/B – 35126, Padova, Italy, fax 39 49 802 1223

Open University for the Environment (Cleon Ricardo dos Santos), Rua Victor Benato 210, Zip 82. 120-110, Curitiba, Brazil, fax 55 41 335 3443

Penang Organic Farm (Lim Poh Im), 116, Jalan Bunga Raya,11700, Gelugor, Penang, Malaysia

Planet Drum, Box 31251, San Francisco, CA 94131, Shatsta Bioregion, USA

Point of View Productions (Karil Daniels, videographer), 3220 Sacramento St., San Francisco, CA 94115, USA. karil@well.com

(Josep Puig i Boix), Sustainable City Councillor, Environment and Urban Services Commission, the Ajuntament de Barcelona, Ciutat 4, 08002 Barcelona, Catalonya, Spain, fax 34 3 4027625

Regenerative Agricultural Resource Center/Rodale Institute (Amadou Diop), BP 237, Thies, Senegal, Fax 22151 16 70, amadou.inbox@parti.inforum.org

RENACE, Chilean Ecological Action Network, Seminario 774 Nunoa, Santiago, Chile

The Resource Exchange (Robin Standish), 636 Colusa, Berkeley, CA 94707, USA, robin@c2.com

Resource Renewal Institute (Green Plans), Fort Mason Center, Building A, San Francisco, CA 94123, USA, fax 1 (415) 945-1654, info@rri.org

Save the San Francisco Bay Association, 1736 Franklin, Oakland, CA 94612, USA, fax 1 (510) 452-9266

Solar Connections (Donna Norris), P.O. Box 2266, Fort Bragg, CA 95437, USA, fax 1 (707) 961-0614

(Joseph Smyth), P.O. Box 1435 Sedona, Arizona 86339-1435, USA

(Zenon Stepniowski), Institute of Architecture and Spatial Planning, Polish Academy of Sciences, Rzeszowkwa 8, 60-468 Poznan, Poland, fax 48-61 663826

(Seydina Issa Sylla), Conservateur-Commandant, National Parks System, Ministry of Environment and Protection of Nature, BP 4055 Dakar, Senegal

Trust for Public Land, 116 New Montgomery St., San Francisco, CA 94105, fax 1 (415) 495-4103, mailbox@tpl.org

Urban Ecology Australia (Paul Downton), 83 Halifax St., Adelaide, Tandanya Bioregion, SA 5000, Australia, fax 61 8 232 4866, urbanec@dove.mtx.net.au

USAID, United States Agency for International Development (Marion Pratt),2200 46th St., NW, Washington, DC 20007, USA

(Michael Vandeman), 3025 Bateman St., Berkeley, CA 94705, USA, mjvande@pacbel.net

(Mariano Vazquez), Polytechnic University of Madrid, Aguilar Del Rio 7, Madrid 28040, Spain, fax 34 1 5975010

(Andrew Venter), consulting biologist, Kruger National Park, P.O. Box 330, St. Lucia 3936, South Africa

Waitakere City, (a city officially striving to be an ecocity), 1 Private Bag 933109, Waitakere City, New Zealand

(Rusong Wang), Department of systems Ecology, Research Center for Ecology and Environmental Sciences, Chinese Academy of Sciences, 19 Zhogguancun Rd., Beijing 100080 china

We The People, 220 Harrison St., Oakland, CA 94607, USA, fax 1 (510) 836-8797, wtp@sirius.com

The Wildlands Project, P.O. Box 1276 McMinnville, OR 97128, USA

(Marian Zeitlin), Tufts University, School of Nutrition Science and Policy, 126 Curtis St., Medford, MA 02155, USA, mzeitlin@infonet.tufts.edu

# Books

ARCOLOGY – THE CITY IN THE IMAGE OF MAN, Paolo Soleri, 1969, MIT Press, Cambridge

AUTOKIND VS. MANKIND, Kenneth Schneider, 1971, W.W. Norton, New York

BORNEO: CHANGE AND DEVELOPMENT, M. Cleary and P. Eaton, 1992, OUP, Singapore

BEYOND SPRAWL – NEW PATTERNS OF GROWTH TO FIT THE NEW CALIFORNIA, Bank of America, California Resources Agency, Greenbelt Alliance, and The Low Income Housing Fund, 1995, Bank of America, San Francisco, California

BOUNDARIES OF HOME – MAPPING FOR LOCAL EMPOWERMENT, Doug Aberly, editor, 1993, New Society Publishers, Gabriola Island, British Columbia

Karil Daniels

Conferee Kate Dorman, foreground, listens to one of the three simultaneously translated languages: Wolof, French or English.

BUILDING A WIN-WIN WORLD, Hazel Henderson, 1996, Berrett-Koehler Publishers, San Francisco

THE CASE AGAINST THE GLOBAL ECONOMY, Jerry Mander and Edward Goldsmith, Eds., 1996, Sierra Club Books, San Francisco

CHANGING THE BOUNDARIES – WOMEN-CENTERED PERSPECTIVES ON POPULATION AND THE ENVIRONMENT, Janice Jiggins, 1994, Island Press, Washington, DC

CITIES AND THE WEALTH OF NATIONS, Jane Jacobs,1984, Random House, New York

COHOUSING, Charles Durrett and Katherine McCamant, 1988, Ten Speed Press, Berkeley, California

COMMON EFFORTS IN THE DEVELOPMENT OF RURAL SARAWAK, MALAYSIA, B. G. Grupstra, 1976, Van Gorcum, Amsterdam, the Netherlands

COMMUNITAS – WAYS OF LIVELIHOOD AND MEANS OF LIFE, Paul and Percival Goodman, 1947, Random House, New York

COMPACT CITIES, Subcommittee on the City, Committee on Banking, Finance and Urban Affairs, US House of Representatives, 1980, US Government Printing Office, Washington, DC

THE COST OF SPRAWL, the Real Estate Research Corporation, for the United States Council on Environmental Quality, Department of Housing and Urban Development and Environmental Protection Agency, 1974, US Government Printing Office, Washington, DC

CREATING ALTERNATIVE FUTURES, Hazel Henderson, 1978, Berkeley Windhover, New York

DESIGN WITH NATURE, Ian McHarg, 1969, Doubleday and Company, Garden City, New York

ECOCITY BERKELEY – BUILDING CITIES FOR A HEALTHY FUTURE, Richard Register, 1987, North Atlantic Books, Berkeley, California

THE ECONOMY OF CITIES, Jane Jacobs, 1969, Random House, New York

ECOLOGICAL DESIGN, Sim Van der Ryn and Stuart Cowan, 1996, Island Press, Washington, DC,

ECOTOPIA, Ernest Callenbach, 1975, Banyan Tree Books, Berkeley, California

EXTINCTION: THE CAUSES AND CONSEQUENCES OF THE DISSAPPEARANCES OF SPECIES, Paul and Anne Ehrlich, 1992, Random House, New York

FEMALE AND MALE IN WEST AFRICA, Christine Oppong, editor, 1983,George, Allen and Unwin, London, chapter by Emile Vercruijsse, "Fishmongers, Big Dealers and Fishermen: Co-operation and Conflict Between the Sexes"

THE FIRST INTERNATIONAL ECOCITY CONFERENCE, THE CONFERENCE REPORT OF, Chris Canfield, editor, 1990, Urban Ecology, Berkeley, California, available from Ecocity Builders upon special request – phone 1 (510) 649-1817

FROM ECOCITIES TO LIVING MACHINES – PRINCIPLES OF ECOLOGICAL DESIGN, Nancy Jack Todd and John Todd, 1994, North Atlantic Books, Berkeley, California

FUTURES BY DESIGN, Doug Aberley, editor, New Society Publishers, 1994, Gabriola Island, British Columbia

THE GAIA ATLAS OF CITIES – NEW DIRECTIONS FOR SUSTAINABLE URBAN LIVING, Herbert Giardet, 1992, Gaia Books Ltd., London, England

GARDEN CITIES OF TO-MORROW, Ebenezer Howard, (originally 1898) MIT Press, 1965, Cambridge, Massachusetts

THE GEOGRAPHY OF NOWHERE, James Howard Kunstler, 1993, Simon and Schuster, New York

THE GRANITE GARDEN, Anne Whiston Spirn, 1984, Basic Books, New York

GREEN PLANS, Huey D. Johnson, 1995, Nebraska University Press, Lincoln, Nebraska

THE DREAM OF THE EARTH, Thomas Berry, 1988, Sierra Club Books, San Francisco

THE IMPACT OF SPECIES CHANGES IN AFRICAN LAKES, Tony Pitcher and Paul Hart, editors, 1996, Chapman and Hall, London

A LANDSCAPE FOR HUMANS, Peter Van Dresser, 1972, The Lightning Tree, Santa Fe, New Mexico

LISTENING TO THE LAND: CONVERSATIONS ABOUT NATURE, CULTURE, AND EROS, Derick Jensen, editor, Sierra Club Books, 1995, San Francisco

NATURAL SYSTEMS FOR WASTE MANAGEMENT AND TREATMENT, S.C. Reed, R. W. Crites and E.J. Middlebrooks, 1995, McGraw-Hill, New York

ÖKOPOLIS, Rudiger Lutz, 1987, Knaur Nachf., München, Germany

ON THE NATURE OF CITIES, Kenneth Schneider, 1979, Jossey-Bass, San Francisco

OUR COMMON FUTURE, The World Commission on Environment and Development, 1987, Oxford Press, Oxford, England

OVERSHOOT: THE ECOLOGICAL BASIS OF REVOLUTIONARY CHANGE, William Catton, Jr., 1982, Urbana, Urbana, Illinois, USA

A PATTERN LANGUAGE, Christopher Alexander, 1977, Oxford University Press, New York

PERMACULTURE ONE, and PERMACULTURE TWO, Bill Mollison, 1978 and 1979, Tagari, Stanley, Tasmania

POPULATION POLICIES RECONSIDERED – HEALTH, EMPOWERMENT AND RIGHTS, Amartya Sen, 1994, Harvard University Press, Cambridge, Massachusetts

THE POWER OF PLACE, Winifred Gallagher, 1993, Poseidon Press, NY, 1993

REBUILDING COMMUNITY IN AMERICA, Ken Norwood and Kathleen Smith, 1995, Shared Living Resource Center, Berkeley, California

SAVING NATURE'S LEGACY: PROTECTING AND RESTORING BIODIVERSITY, Reed F. Noss and Allen Y. Copperrider, 1994, Island Press, Washington, D.C.

SEEDS FOR CHANGE – CREATIVELY CONFRONTING THE ENERGY CRISIS, Deborah White et. al, 1978, Dominion Press, Blackburn, Victoria, Australia

SEX AND VIOLENCE: ISSUES IN REPRESENTATION AND EXPERIENCE, Ed. P. Harvey and P. Gow, 1994, Routledge, London; chapter by Henrietta Moore, "The Problem of Expanding Violence in the Social Sciences"

SUSTAINABLE CITIES, Bob Walter, Lois Arkin and Richard Crenshaw, editors, 1992, Eco-Home Media, Los Angeles

TOWARDS AN ECO-CITY – CALMING THE TRAFFIC, David Engwicht, 1992, Envirobook, Sydney, Australia

TOWNSCAPE, Cullen, Gordon, 1961, The Architectural Press, London

THE UNIVERSE STORY, Brian Swimme and Thomas Berry, 1992, Harper Collisn, 1992, New York

URBAN ECOLOGICAL DEVELOPMENT: RESEARCH AND APPLICATION, Rusong Wang and Yonglong Lu, editors, 1994, China Environmental Science Press, Beijing

USING RESIDENTIAL PATTERNS AND TRANSIT TO DECREASE AUTO DEPENDENCE AND COSTS, John Holtzclaw, 1994, Natural Resources Defense Council, San Francisco

THE WEB OF LIFE – A NEW SCIENTIFIC UNDERSTANDING OF LIVING SYSTEMS, Fritjof Capra, 1996, Doubleday Books, New York

WINNING BACK THE CITIES, Peter Newman and Jeff Kenworthy, with Les Robinson, 1992, Australian Consumers Association, Marrikville, New South Wales, Australia and Pluto Press, Leichhardt, New South Wales, Australia

WORLD AGRICULTURE: TOWARDS 2010, Nikos Alexandratos, Editor, 1995, Food And Agriculture Organization of the United Nations, New York

THE WRIGHT SPACE, Grant Hildebrand, 1991, University of Washington Press, Seattle, Washington, USA

# Periodicals, Articles and Speaker-Recommended Papers

Auto-Free Times, a publication of the Fossil Fuels Policy Action Institute and the Paving Moratorium, P.O. Box 4347, Arcata, CA 95518, USA, phone 1 (707) 826-7775, $30US/year

Cohousing, the Journal of the Cohousing Network, P.O. Box2584, Berkeley, CA 94702, coho@aol.com, $20US/year

Communities Magazine – Journal of Cooperative Living, P.O. Box 169, Masonville, CO 80541-0l69, USA, $18US/year

"Constructed Wetlands for Wastewater Treatment and Wildlife Habitat - 17 Case Studies," United States Environmental Protection Agency (EPA), 1993, EPA R-93-005, Washington, D.C., September

The Design eXchange for Sustainable Living, Ecovillage Training Center, Ecovillage Network of the Americas Global Ecovillage Network, 560 Farm Road, P.O. Box 90, Summertown, TN 38483-0090, USA, ecovillage@thefarm.org, $25US/year

The E. F. Schumacher Society Newsletter, 140 Jug End Road, Great Barrington, MA 01230 USA, fax 1 (413) 528-4472, efssociety@aol.com

Earth Island Journal, Earth Island Institute, 300 Broadway, Suite 28, San Francisco, CA 94133, USA. earthisland@igc.apc.org, $25US/year

The Ecocity Builder, Ecocity Builders, 5427 Telegraph Ave., W2, Oakland, CA 947609, USA, ecocity@igc.qpc.org, $20US/year

Ecocity Cleveland, 2841 Scarborough Road, Cleveland Heights, OH 44118, USA, $20/year

Fremontia – Journal of the California Native Plant Society, 1722 J St., Suite 17, Sacramento, CA 95814, USA, $35US/year

"The Environmental Consequences of Having a Baby in the United States," Charles Hall with R. G. Pontius, L. Coleman and J. Y. Ko,Wild Earth (magazine), summer, 1995

"Human Appropriation of Fresh Water," Sandra Postel, Paul Ehrlich and G.C. Daily, Science (magazine), February 9, 1996

"Human Appropriation of the Products of Photosynthesis," Peter Vistousek, Paul and Ann Ehrlich, and P. A. Matson, Bioscience (magazine), vol. 36, pages 368 to 373, 1986.

IAA Review, International Academy of Architecure, 2 Tzar Osvoboditel Blvd., Sofia, Bulgaria, fax 35 92 9814945

"Inner City Regeneration – making it happen," Greg Freeson, Housing Review (magazine), the Housing Centre Trust, London, England, February, 1995

New York Streetcar News, Committee for Better Transit, Inc., P.O. Box 3106 Long Island City, NY 1103, $12US/year

Public Innovations Abroad, The International Academy for State and Local Government, 444 North Capitol St. N.W., Suite 345, Washington, DC, 20001, fax 1 (202) 434-4851, $48US/year

The Permaculture Activist, P.O. Box 1209, Black Mountain, NC, 28711, USA, $19US/year

Sierra Magazine, Sierra Club, 85 2nd Street, 2nd Floor, San Francisco, CA 94105, $25/US)

TRAC, Train Riders Association of California, 926 J Street, Suite 612, Sacramento, CA 94814, USA, $25US/year

The Urban Ecologist, Urban Ecology, 405 14th St., Suite 900, Oakland, California, 94612, USA, $35US/year

"World Map of the Status of Human-Induced Soil Degradation," Leonard Oldeman, R. T. A. Hakkeling and W. G. Sombroek, a United Nations Environmental Program paper, UNEP, Nairobi, Kenya, 1991.

# Index

Yoff, Senegal.